DO YOUR OWN
WIRING

K.E. Armpriester

Sterling Publishing Co., Inc.   New York

Consultants

J.A. Tedesco was formerly an Electrical Field Service Specialist with the National Fire Protection Association, Quincy, MA, and Associate Editor of the National Electrical Code Handbook. Currently, he is an electrical designer/consultant and has extensive experience in inspection and investigation of electrical fires.

Hans Grigo is the Manager of Home Safety for the National Safety Council, Chicago, Illinois. He is also the NSC house and yard expert. The National Safety Council is a public service organization whose mission is to educate and influence society to adopt safety and health policies, practices and procedures.

Martin M. Mintz, AIA is the Director of Technical Services for the National Association of Home Builders, Washington, D.C. He is a registered architect and member of the American Institute of Architects. Mr. Mintz is a lecturer and writer, and serves on several committees including the NAHB Research Committee and a National Institute of Building Sciences committee.

Acknowledgments: Lyons Electrical Supply and Fixtures, Dayton, Ohio; PK Building Center, Englewood, Ohio; Radio Shack, a Division of Tandy Corporation, Dayton, Ohio; Wertz Hardware, West Milton, Ohio

Thanks to the following individuals for the use of photographic locations: Linda Ball, Candace Curry, Susan Goutman, Tina Grismer.

Book design by Linda Watts
Produced by Bookworks, Inc. Illustrations by Mary Kay Baird, Linda Ball, Sylvia Schwartz, and O'Neil & Associates

NEC® and National Electrical Code® are registered trademarks of the National Fire Protection Association, Quincy, Massachusetts.

Library of Congress Cataloging-in-Publication Data

Armpriester, Kate, 1947–
        Do your own wiring / K.E. Armpriester.
                p.   cm.
        Originally published: New York, NY : Popular Science Books, ©1987.
        Includes index.
        ISBN 0-8069-8472-4
        1. Electrical wiring, Interior—Amateurs' manuals.   I. Title.
        [TK3285.A49   1991]
        621.319′24—dc20   91-18970
                                                                                    CIP

10   9   8   7   6   5

Published in paperback 1991 by Sterling Publishing Company, Inc.
387 Park Avenue South, New York, N.Y. 10016
Originally published by Popular Science Books
© 1987 by K. E. Armpriester and Bookworks, Inc.
Distributed in Canada by Sterling Publishing
% Canadian Manda Group, P.O. Box 920, Station U
Toronto, Ontario, Canada M8Z 5P9
Distributed in Great Britain and Europe by Cassell PLC
Villiers House, 41/47 Strand, London WC2N 5JE, England
Distributed in Australia by Capricorn Ltd.
P.O. Box 665, Lane Cove, NSW 2066
Manufactured in the United States of America
All rights reserved

Sterling ISBN 0-8069-8472-4

Contents

Circuitry and Safety

Electricity is something that most people take for granted until it goes out. If you've ever been preparing to dry your hair, iron a shirt, or watch TV when this happens, then you know the frustration. Even worse are those mornings when the alarm doesn't ring because a thief in the night stole the power.

We all have a tremendous dependency on these little things called electrons...and maybe the power electricity has over us also causes us to feel intimidated or frightened by it. It plays such a major role in our lives that we assume it must be complex and difficult.

Yet wiring is one of the least difficult of all the home skills to perform. It requires very little equipment, the equipment is standardized, it creates little or no mess, and the projects take relatively small amounts of time to complete. The only key to correct and safe wiring is understanding what you are doing.

Please read this chapter on circuitry and safety carefully, especially if you are a beginner. It is very important and will help you to understand not just *what* you are doing but *why* you are doing it. If you aren't a beginner you should read this chapter anyway because it will refresh your memory with safety precautions about this potentially dangerous skill.

Circuitry Made Simple

It all begins with the electrical circuit. A *circuit* is a continuous path or a circle. Within your home a circuit is completed when the current, beginning at the fuse box or breaker box, travels through wiring to a fixture such as a light and then returns to the box. In this way your electrical system differs from your water supply system. When you turn on a faucet you consider that the water is leaving your home...escaping. But the electrical current remains in your home through means of the neutral conductors (usually white) and equipment grounding conductors in the system. It is only through the 'hot' wires (usually black or red) that electricity is 'escaping' and emitting a charge.

In another way, however, electrical systems are similar to plumbing systems. Our water travels from a reservoir into our houses. Electricity is pushed to our houses by a generator. So both water and electrical energy have pressure that is causing this motion. And just as we have faucets that release the water, we have outlets in the form of lights, receptacles, and switches that release electrical energy.

An electrical current travels through a conducting material. A *conductor* is anything that permits the flow of electricity rather than resisting it. Copper is a good conductor of electricity; rubber and thermoplastic are good resisting materials. Therefore, these substances are often used in the composition of wiring to, respectively, conduct and contain the flow of electricity.

The conducting element or wiring must be of an adequate size to contain the flow of electricity. If the wires are too small to permit swift passage of the current, they will grow hot and break. This "open circuit" changes the flow and has the potential for dangerous consequences.

A Helpful Equation. Electrical current is defined in measurable terms called *watts, volts,* and *amperes.* Watts is perhaps the most familiar term since we have all purchased light bulbs according to their watts rating. The watts measurement refers to the rate at which a device consumes energy. Amperes or amps are units that measure the amount of electrons passing, per second, through a given point on a circuit. The pressure which causes the current to flow is measured in units called volts.

Thus the standard electrical formula computes the pressure times the amount of electrons:

VOLTS X AMPS = WATTS

An understanding of this equation will help you to figure a load on a circuit, plus it will aid you when purchasing appliances and when purchasing materials for doing your own home wiring.

For example, if you wanted to purchase an air conditioner but weren't sure if your present circuit could easily accept it, you could add its amp rating, 8, to the total amps already being used on the circuit. (Large appliances are rated according to amps but smaller ones such as light bulbs are rated by their wattage. In order to convert the smaller appliances' usage to amps simply divide the watts by the volts.) In this case the total watts from all other appliances on this circuit is 400 and the circuit is 120 volts. Four hundred watts divided by 120 volts equals 3.3 amps and that added to the air conditioner's rating of 8 equals 11.3 amps.

The circuit breaker has 15 amps listed as the circuit's capacity so you've determined that adding your air conditioner now brings your total amps to 11.3 of the 15 that the circuit can hold. Such equipment should not exceed 80 percent of the total, so 12 amps is the maximum on this circuit.

Codes and Permits

Electricity can be dangerous and that's why it is tightly regulated by national and local codes. The *National Electrical Code®* (often called the *NEC®* or simply the 'Code') is a volume of rules published by the National Fire Protection Association, a non-profit organization. These rules take into account practically every conceivable electrical situation, and they are updated every three years. Though not strictly laws, this set of regulations serves as a model for the laws of local communities. Some counties or towns simply adopt the *NEC* while others adopt their own set of codes based on it.

The replacement and repair projects in this book rarely require permits. However, if you are extending an existing circuit or adding a new one, you might need a permit and you should contact your community's building department to obtain one. You might also be required to have your work checked by a local electrical inspector or licensed electrician.

Your Utility Company. If you intend to make any changes in wiring that will affect your service, you should contact your utility company. The most notable example would be to go from a 120-volt 2-wire service to a 120/240-volt 3-wire service. Another situation that warrants calling your local power company is doing building or remodeling that changes the location of your service entrance panel or meter.

All About Grounding

Electricity always follows the path of least resistance and this principle is essential to understanding *grounding*. Grounding basically means directing a live current into the ground, thus making it harmless. Copper wiring is more conductive than our own bodies to a flow of current and that's why it is possible to touch something that is electrified and not get shocked—*but only if it is properly grounded*.

Two hazards can result from ungrounded circuits—fire and shock. A fire can begin as a result of a *short circuit* (a 'shortcut' that the circuit takes instead of traveling its normal path). If a short circuit develops from a broken hot wire, such as across worn wiring insulation, then heat is generated—heat which could eventually cause a fire. Because the path offers high resistance, the circuit breaker or fuse will not be tripped.

Shocks can be caused by abnormal current flow resulting from hot wires that are touching conductive materials, such as power tools and appliances with metal housings; metal switches, junction and outlet boxes; and metal faceplates.

But grounding assures us that these hazards will not occur. A special set of wires leads the current back to the ground. And because a leaky wire or an abnormal current generates increased amperage on the way to ground, the current will trip off a circuit breaker or blow a fuse, and thus shut down the entire circuit.

UNGROUNDED AND GROUNDED CIRCUITS

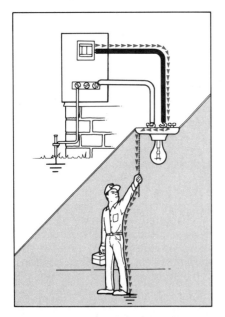

An Ungrounded Circuit. In this illustration, the energized conductor has accidently slipped off the terminal causing the fixture and the pull chain to become 'hot' or electrically charged. Touching the chain under these conditions creates a path to ground for the current and consequently the person would receive an electric shock.

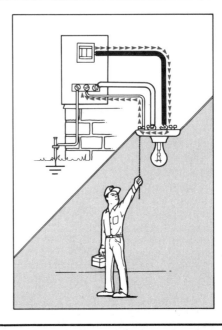

A Grounded Circuit. Here, the circuit has a grounding system in the form of an equipment grounding conductor (often bare copper) that connects the metal housing of the light fixture to the neutral bus bar. This provides an optional path to ground in the event of a ground fault. Danger of being shocked is minimized plus, because the current is unusually high, a circuit breaker trips or a fuse blows, and the entire circuit shuts down.

Methods of Grounding. Generally, most homes are grounded through their plumbing systems by means of a buried metal water pipe that is at least 10 feet long. Yet some homes do not have the appropriate plumbing systems to adequately accept a ground wire. In this case, an alternate grounding system should be used. It incorporates one or two copper rods or ¾″ galvanized pipe driven at least 8 feet into the ground.

A Grounding Electrode System. A grounding system needs only metal conductors to operate prope‍~~~~~ but with today's use of non-me~~~~~~~~~pipe and insulated fittin~~~~~~~~~~uld be broken. Oth~~~~~~~~~~~~~res that could rende~~~~~~~~~~~ ineffective inclu~~~~~~~~~~ in your water ~~~~~~ from the meter t~~~~~

Because of this, the Code now requires the *grounding electrode system,* one of three kinds of supplementary systems described below. If you are installing new wiring or upgrading old wiring, you should ensure proper grounding by using this system. If you presently have a No. 6 copper grounding electrode conductor connected to a water pipe buried 10 feet or more underground, you must supplement that system in one of the following ways:

■ Connect the grounding electrode conductor from the service entrance panel to a No. 2, or larger, bare copper wire at least 20 feet long that is buried 2½ feet deep alongside your house.

■ Connect the grounding electrode conductor to 20 feet or more of ½-inch steel reinforcing rod or No. 4, or larger, bare copper wire that is enclosed in concrete near the bottom of the concrete foundation of your house.

■ Connect the grounding electrode conductor to the metal casing of your well if you have one (but not the drop pipe in the well).

Local codes may vary in regard to these procedures or may stipulate that you take even further precautions. In any case, the grounding electrode system is very important.

Grounded Outlets. In new homes, all receptacles have three-slot grounded outlets. But many older homes still have outlets with only two slots. This sometimes means that the outlet is not grounded. In order to check for grounding, simply shut off the current to the receptacle (page 5), remove the faceplate and inspect the box. If there is no bare copper grounding wire attached to the switch or box, or no BX cable present (page 85), the system is not grounded.

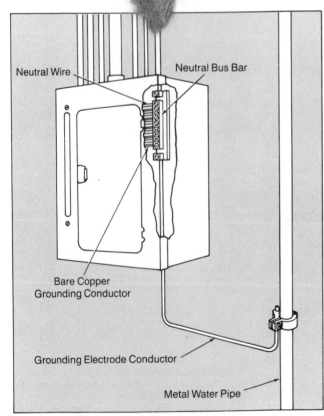

Neutral Wire

Neutral Bus Bar

Bare Copper
Grounding Conductor

Grounding Electrode Conductor

Metal Water Pipe

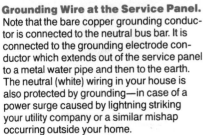

Grounding Wire at the Service Panel.
Note that the bare copper grounding conductor is connected to the neutral bus bar. It is connected to the grounding electrode conductor which extends out of the service panel to a metal water pipe and then to the earth. The neutral (white) wiring in your house is also protected by grounding—in case of a power surge caused by lightning striking your utility company or a similar mishap occurring outside your home.

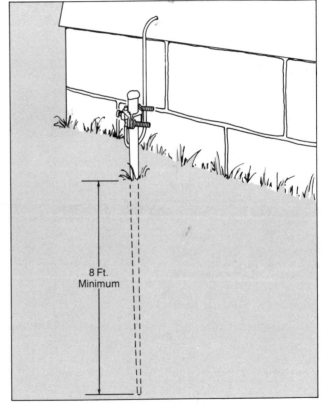

8 Ft.
Minimum

An Alternate Ground. When there is no buried water pipe at least 10 feet long, the grounding electrode system consists of a ⅝″ copper-coated or steel rod or ¾″ galvanized pipe driven 8 feet or more into the earth. In regions where the earth is very dry, two rods may be necessary. Occasional checks of their resistance by an electrician are also advisable, as dryness or corrosion from acid soil could lessen the efficiency of the rod(s).

If the system is grounded, then you can either replace this receptacle and all of your two-slot receptacles with new three-slot receptacles or you can use a *grounding-adapter plug*. This device is designed with a rigid ear or lug, usually green in color, that fits beneath the cover plate screw and serves to ground the three-slot adapter.

Ground-Fault Circuit Interrupters. Even if your receptacle is properly grounded you can still receive a shock from a very minor circuit leakage, one that is too small to trip a circuit breaker or to blow a fuse. This scenario, which can be deadly, occurs in damp or wet locations inside and outside your home. That is why the Code now requires *ground-fault circuit interrupters* in all high-risk areas of new homes, such as bathrooms, basements, and garages as well as outdoor grade-level outlets and kitchen outlets within 6 feet of the sink.

These sensitive devices compare the amperage entering a fixture on the black hot wire to the amperage leaving on the white neutral wire. If the GFCI detects as little as .005 amp difference between currents, it breaks the current in 1/40 of a second, preventing dangerous shock. Though they might not be required in laundry areas or workshops yet, they should be seriously considered in any potentially wet areas. You can install them as outlets or you can protect your entire system by installing one in your service panel or subpanel.

Safety, The Most Important Issue

By now you have surely gained a respect for electricity by learning all about circuits and grounding. But what follows here is more important than anything else in this book so please read carefully and memorize these precautions.

WARNING

■ Before working on your wiring, ALWAYS TURN OFF THE POWER.

■ Before touching any wire, ALWAYS TEST WITH A VOLTAGE TESTER (page 14) TO MAKE SURE THAT THE POWER IS OFF.

Using a Grounding-Adapter Plug. First, make sure that the cover plate metal screw is properly grounded by checking it with a voltage tester (page 14). Next, remove the screw, plug in the adapter, with the ear over the screw hole, then replace the screw. Cover plate screws are often short so you should check to make sure that the screw is grounded before plugging in the adapter. If not, replace it with a longer (6/32) screw.

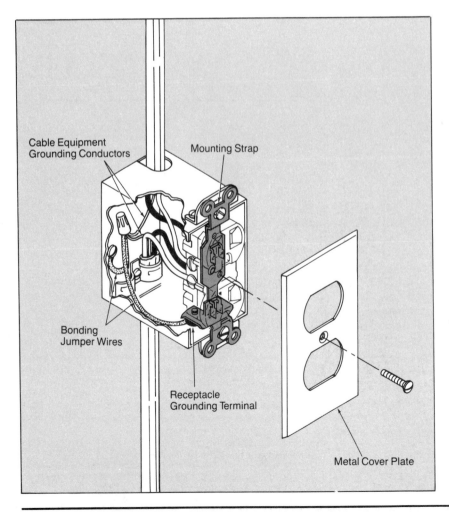

Cable Equipment Grounding Conductors

Mounting Strap

Bonding Jumper Wires

Receptacle Grounding Terminal

Metal Cover Plate

A Grounded Receptacle. If something went wrong, the metal parts of this receptacle would not dangerously conduct current because it would be directed to ground by the bare copper wires of the cable. The metal strip running through the inside of the receptacle (the mounting strap) is connected to the entire box by screws, plus the receptacle cover plate, if metal, is connected by a screw. All of these metal parts are in contact with the receptacle grounding terminal. This, plus the cable equipment grounding conductors, are connected to the box grounding screw by way of bonding jumper wires, providing a safe pathway for current.

In addition to these two most important rules, here are others that you should learn for your own protection:

1 **Turn off the breaker or remove the fuse** of the circuit you are working on. Then post a sign on the panel so that no one will turn it back on while you're working on it. To be extra safe, padlock the panel door and keep the key with you until you're ready to turn the power back on.

2 **Don't stand on a damp floor** or allow any part of your body to be damp when working with electricity. If the floor around your service panel is even occasionally damp, keep some boards or a rubber mat there to stand on while touching the box.

3 **Never touch any plumbing or gas pipes,** metal ductwork or any other metal object when working with electricity; pipes are considered as grounded objects. Even the metal shell of a grounded electric drill can be hazardous. Touching a hot wire and a pipe at the same time could cause current to flow through your body. Therefore, you should be careful to have your balance and avoid using your free hand to steady yourself if you slip.

4 **To protect your eyes, wear safety glasses or goggles.** This precaution should be taken when testing at outlets in case of an electrical arc and also when doing construction or dismantling that involves filing, drilling, or sawing.

5 **After completing your work, turn on the power and test it** with a voltage tester (page 15). The tester should light when a connection is made between the grounded box and the black hot wire. It should not light when a connection is made between the grounded box and the white neutral wire.

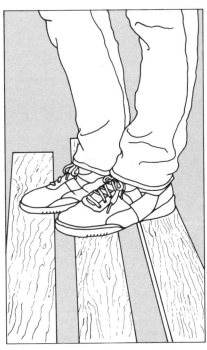

Guard Against Damp Floors. When performing electrical work on damp or potentially damp floors, always stand on a dry rubber mat or on dry boards.

GROUND-FAULT CIRCUIT INTERRUPTERS

1 **Built-in GFCI Receptacle.** Easy to install (see page 99 for instructions). If installed as the first receptacle on a circuit, it will protect the other receptacles further along the circuit.

2 **GFCI Breaker.** For insertion into the service panel or subpanel. Available for 120-volt and 240-volt circuits. A licensed electrician should install this device.

Wiring Tools and Materials

f you were to go to an electrical supply store, your mind might be boggled by all of the different kinds of equipment on display; it would be a shock of another kind! But for the home wiring skills presented in this book, you'll need only a meager amount of tools and materials—especially if you're only interested in making repairs.

In this chapter, you'll learn about all the tools and materials you will have to acquire to develop basic wiring skills. You will discover that there are specific tools for wiring that make the job extra easy—such as a multipurpose tool that is actually several tools in one.

But after you've acquainted yourself with these items, please don't rush out and buy all of them. First, determine what your project or projects are going to be. Then, make your shopping list.

Since the amount of tools and materials is relatively small compared to other home skills, we recommend that you purchase only high-quality items. If you've ever used cheaply made supplies, then you know how easily they can break and what an aggravation it is to have to run to a hardware store in the middle of your project to replace something.

Good tools are always a wise investment. They can be used over and over again...and many of the tools, such as screwdrivers and hacksaws, will be used for other home skills. Besides, once you learn these skills, you'll probably want to do more wiring.

Basic Wiring Tools

You can do a great deal of wiring with nothing more than a pair of pliers and a screwdriver. But in order to make your projects flow as smoothly and easily as possible, you should invest in some, if not all, of the following tools. There are specialized electrician's tools that will save you time and headaches—for stripping and bending wire and for drilling and cutting through walls.

Some of these tools contain extra features that make them a little safer for wiring—such as lineman's pliers that have special insulated handles as an extra measure against shock. So even if you have tools in your tool kit right now that you think might be useful in wiring projects, you might want to inspect them carefully to make sure that they're similar to what we show here.

On the other hand, there will be tools used for some of the projects in this book that are so standard that we've elected not to describe them here. These include an electric drill, tape measure, flashlight, hammer, saws, chisels, and screwdrivers. And though it also isn't shown here, you should invest in a tool pouch so that you can quickly and easily find your tools when you need them.

There are also other tools that are not shown here that might be shown or referred to elsewhere in the book. For example, the conduit bender is used only for working with rigid metal or thinwall conduit in outdoor wiring and is therefore introduced in Chapter 11.

UL Listed Materials

Underwriters' Laboratories in the United States and the Canadian Standards Association in Canada are independent, non-profit organizations whose purpose is testing electrical equipment to determine its quality. Therefore, the term 'UL listed' means that a particular material or device is listed or labeled for general use by the UL organization.

TOOLS FOR HOME WIRING

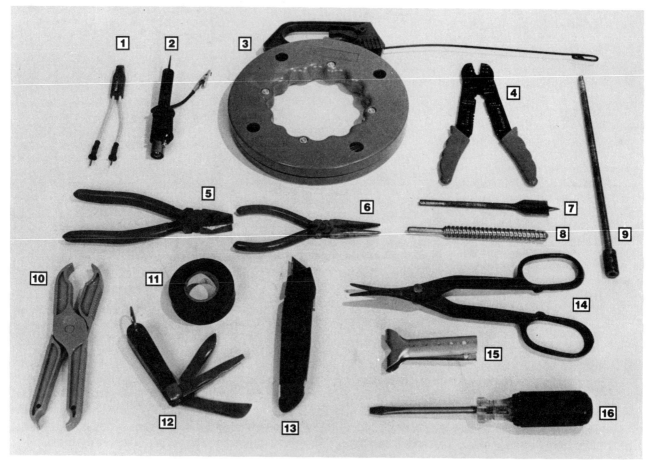

1 Voltage Tester. A wiring essential. Composed of a neon bulb and two insulated wires that end in metal probes. Used to check that the current is off before you begin a job. Also used with power on to test for proper grounding and, in special cases, to check that voltage is available in wires. When buying, be sure to get a standard (90-500) voltage tester and not a low-voltage one.

2 Continuity Tester. Operates with a small battery. Used only with the power off to pinpoint wiring malfunctions in components such as broken light sockets or switches. Also used to check continuity of conductors in entire circuits.

3 Fish Tape. A must whenever you're going to pull wires or cable behind walls, through conduit or raceways. Long fish tapes come wound around a metal reel; short ones come without a reel. Note: You should purchase two of these in appropriate sizes for your job.

4 Multipurpose Tool. Combines as many as seven different functions, the most important being cutting wires and stripping insulation from them. Wire gauges indicate which hole to use for which size wire. May also be used for cutting small bolts and crimping special kinds of wire connectors. Shop carefully for this, as some designs have weak points.

5 Lineman's Pliers. Includes a wire cutter and striated jaws. May also be used for pulling wire, bending heavy wire, and twisting out removable parts of certain electrical components.

6 Needlenose Pliers. An essential. Also called long-nose pliers; may have a wire cutter near the pivot. Necessary for curling wires into the loops needed for electrical connections.

7 Spade Bit. For attaching to your electric drill. Though other size bits may be needed for special projects, a ¾-inch bit will suffice for the holes in wood through which cable can be run.

8 Masonry Bit. For holes through masonry, use a ½-inch carbide-tipped masonry bit—the largest that will fit on a ⅜-inch drill.

9 Bit Extension. An extension that will allow you to drill through thick or widely spaced beams.

10 Fuse Puller. A plastic tool that makes it safe to remove fuses without the danger of electrical shock.

11 Electrician's Tape. Sometimes used for identifying white neutral wires as black or hot wires. Also may be used to secure a connection made with a wire nut. Formerly used for splicing, tape has been generally replaced by wire nuts although it is still permitted by code for this practice.

12 Electrician's Knife. A special multipurpose tool that includes a knife for stripping off nonmetallic cable.

13 Utility Knife. For cutting through drywall when making wiring runs and installing boxes.

14 Metal Shears. Also called 'aviation shears' or 'snips', these may be used for trimming flanges off of certain components. When buying, select straight instead of curving blades.

15 Cable Ripper. An inexpensive helper. Rips the outer insulation sheath of flat nonmetallic 2-wire cable, with or without grounding wire.

16 Electrician's Screwdriver. Different in design than a regular screwdriver. Has a long slender blade with the flat portion of the tip the same width as the round shaft. Allows you to work in tight spots and drive screws that are deeply recessed.

ELECTRICIAN'S TIP: The use of insulated tools is recommended—for safety and ease of handling. But don't depend on tools for safety... always work with the circuit shut off.

When purchasing materials be sure to look for this marking—it is your assurance that your purchase will safely perform the function for which it was intended. For example, a UL listed lamp cord will work fine on that fixture, but it should never be used as wiring for the home.

Conductors—Wires and Cables

Electricity is carried by conductors that range in size from small lamp wires to high-voltage wires that travel across the country. *Wires* are the smallest of these conductors. *Cables* are larger pre-assembled combinations of wire which are grouped together in a protective outer sheath usually made of nonmetallic material.

Wire Sizes and Uses. If a wire or cable is too small for the amperage that is being conducted through it, that conductor will overheat, damage the insulation, create a ground fault or short circuit, and possibly cause a fire. That's why the NEC has stipulated what size wire should be used in any wiring project.

Wire sizing is controlled by the American Wire Gauge System (AWGS) which specifies the size and type to be stamped on the insulation. In wire sizes, the smaller the number, the larger the wire; so a No. 10 would be larger than a No. 12 wire.

Most house wiring for lighting circuits, and many of the projects in this book are done with No. 14 or No. 12 copper wire. If you are installing new wiring, it is advisable to use No. 12 which is more efficient, not that much more expensive, and generally more acceptable in local codes.

Types of Wires. The two most frequently used wires in home circuitry are *Type T* and *Type TW*. Type T, which stands for thermoplastic, the wire's coating material, is able to withstand a wide range of temperatures but should only be used in dry locations. Type TW is weatherproof and is thus recommended for use outdoors and in damp areas such as basements.

Wire Colors for Easy Assembly. The coating on wires is of different standardized colors that indicate how each wire is to be used and what it should be connected to. Some terminal screws are also color-coded to aid you even further. Black and red are most commonly used to indicate hot wires, and white usually indicates neutral. But don't rely solely on these colors—always use a voltage tester to determine if a wire is hot regardless of its color.

COPPER WIRE SIZES AND AMPACITY

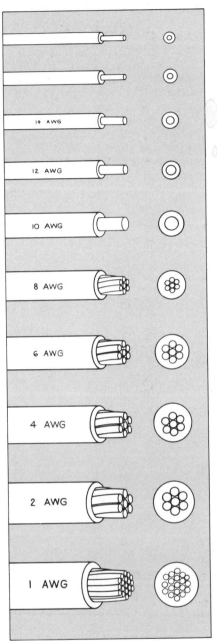

Note that the wire size is printed on all wires except Nos. 16 and 18, which are too small. AWG indicates that the size is based on the American Wire Gauge System.

No. 18—7 Amperes. Used for low-voltage systems like thermostats, doorbells, intercoms, and small appliance wiring. In stranded form.

No. 16—10 Amperes. Same use as No. 18.

No. 14—15 Amperes. Used for standard 120-volt lighting circuits. Also used for circuits that have receptacles for small appliances such as televisions and clocks.

No. 12—20 Amperes. Same use as No. 14. Used for wiring of small appliance receptacles in kitchens, family rooms, dining rooms, and pantries. May also be used for refrigerators.

No. 10—30 Amperes. Used in 120/240-volt circuitry for larger appliances such as clothes dryers, counter-mounted cooking units, and electric water heaters.

No. 8—40 Amperes. Used in 120/240-volt circuits for electric ranges.

No. 6—55 Amperes. Used in 240-volt circuits for appliances such as central air conditioners, heat pumps, electric ranges, and ovens.

No. 4—70 Amperes. Same use as No. 6.

No. 2—95 Amperes. Used for service entrance conductors.

No. 1—110 Amperes. Same use as No. 2.

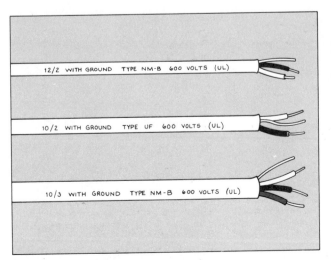

Nonmetallic Cables. Note the printing on these three common types of nonmetallic cables. It identifies, from left to right: the wire size and the number of conductors (these are combined with a slash between them, for example 12/2); the presence of a ground wire (sometimes shown as "WG"); and lastly, the cable type.

Color Code for Wires

Color	Function
Black	Hot Wire
Red	Hot Wire
Blue	Hot Wire
White marked as Black	Hot Wire*
White	Neutral Wire
Green	Grounding Wire
Green & Yellow	Grounding Wire
Bare Copper	Grounding Wire

*White always indicates a neutral wire except when it is marked, usually with black electrician's tape, in a switch loop. See pages 30-31 for a thorough explanation.

Types of Cables. Though there are several types of cables, for your purposes in doing home wiring, there are only three that you need to become familiar with. All three types are *nonmetallic sheathed cable* also referred to as 'plastic sheathed cable' or 'Romex', a popular brand name. These cables all have nonmetallic coverings which make them flexible and easy to use.

NM Cable. This type is used only in dry locations and is often used in house circuits. It is composed of two or three conductors and an equipment grounding conductor (in this book called the grounding wire). The conducting wires are coated with nonmetallic insulation while the ground wire is usually bare. All of the wires are surrounded by a paper insulator which is covered by the outer sheath of moisture-resistant, flame-retardant material.

IMPORTANT

NM cable is available without a grounding wire, but it is highly recommended that you use only cable that contains it.

TYPES OF CONDUIT

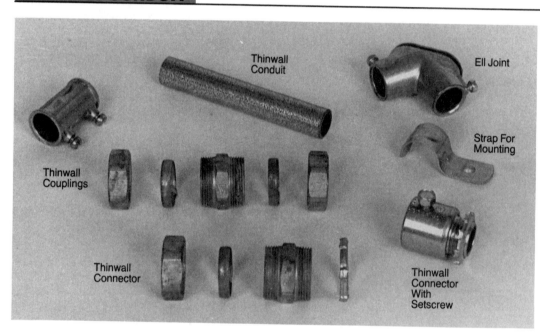

Thinwall Conduit

Ell Joint

Strap For Mounting

Thinwall Couplings

Thinwall Connector

Thinwall Connector With Setscrew

Thinwall Conduit (EMT). Thinwall conduit, called *EMT* for *electrical metallic tubing*, is usually made of galvanized steel. It protects wires from physical damage better than plastic-sheathed cable but is difficult to install because of its inflexibility.

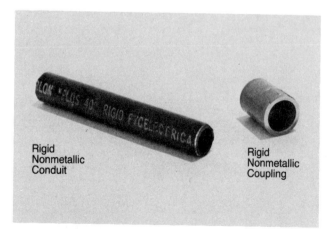

Rigid Nonmetallic Conduit

Rigid Nonmetallic Coupling

Rigid Nonmetallic Conduit (PVC). Rigid nonmetallic conduit is very resistant to corrosion but it is more expensive than metal conduit. Assembly with adhesive is easy, but once joined, it cannot be taken apart.

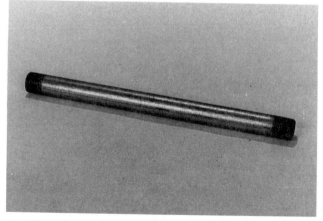

Rigid Metal Conduit. Also called 'heavy-wall' conduit, this type is available in either galvanized steel or aluminum. Can be cut with a hacksaw and bent with a conduit bender. See page 99 for components.

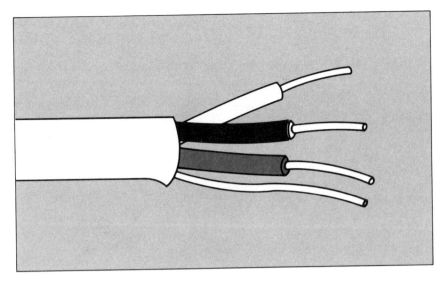

Anatomy of a Cable. This nonmetallic sheathed cable, Type NM, is commonly used in home wiring. Note that it contains three conductors—a white neutral wire and black and red hot wires. These are thermoplastic coated wires which, if they touched each other, could cause a short circuit. Type NM also is required to contain a bare equipment grounding conductor (called the *grounding wire* in this book) which, like the other wires, is wrapped in paper, a moisture-resistant substance. The entire cable is enclosed in flexible nonmetallic material.

ELECTRICIAN'S TIP: Be sure to use the new NM-B type cable now required by the Code. It has 90° C conductor insulation.

NMC Cable. Similar to NM cable but especially designed for use in damp areas such as basements and outdoors. The interior of this cable reveals that all of the wires are imbedded in a solid nonmetallic sheath. Not available in some areas; as a substitute you should use UF cable.

UF Cable. UF stands for 'underground feeder'. Waterproof, this cable can be buried outdoors with no additional protection. With its flexibility it's much easier to work with than conduit. It can also be used in outbuildings that are subject to excessive moisture.

New MC Cable. Yet a fourth type of cable is now available these days. Called MC or 'Ready/Clad' (a brand name), this new type of cable is sheathed in aluminum instead of nonmetallic material. It's slimmer than nonmetallic types and is easier and faster to use. MC cable may be used in exposed or concealed locations, but if it is buried in the ground or cast within concrete it should be a PVC-coated variety.

Conduit—Protective Coverings for Conductors

Conduit is piping of plastic or metal through which wiring is run after it has been installed. It protects conductors from moisture and physical harm. Thinwall conduit (EMT) is occasionally used within homes as is ENT; the other two types are typically used in outdoor wiring.

Thinwall Conduit. Also called EMT, this galvanized steel or aluminum tubing is joined by set screw or compression-type couplings that exert enough pressure to hold lengths together. It should be cut with a hacksaw and, once cut, rough edges must be smoothed with a half round file.

New ENT Conduit. This new product, actually flexible PVC conduit, offers some advantages over EMT. It bends easily and installs in shorter time. Typically blue in color, ENT accepts TW wires. Special ENT boxes are available for connections to the conduit or it may be connected to regular metal boxes with the use of ENT box terminators. ENT can be cut with a plastic pipe cutter which leaves no burrs, or with a knife, in which case the burrs must be removed.

Rigid Nonmetallic Conduit. Commonly called PVC, it is very easy to work with because it is lightweight and is joined with a special glue that sets in seconds. It should also be cut with a hacksaw and rough edges must be scraped away with a knife blade. This conduit is impossible to take apart once glued.

Rigid Metal Conduit. This metal conduit, also known as 'heavy-wall' conduit, is heavier than the thinwall variety and is connected by threaded joints. It can be cut with a hacksaw and must be bent with a special tool called a *conduit bender.*

A WORD ABOUT ALUMINUM WIRING

Aluminum wiring in itself is not dangerous but because it expands and contracts more than copper wire, it can work loose from a terminal and cause a fire. Special care must be taken to install aluminum wiring safely and correctly and the manufacturer's directions must be followed specifically. Therefore, in this book, all of the projects are shown only with copper wiring.

You should check your system for aluminum wiring, however. If your residence was built prior to 1965 and has had no additional or replacement wiring since that date, there is little chance that aluminum was used.

If your home was built after 1965 or has had additional wiring since then, you should contact the builder or electrical contractor to determine if aluminum wiring was used. **DO NOT ATTEMPT TO MAKE THIS DETERMINATION ON YOUR OWN.** If aluminum wiring was used, you should have a licensed electrician or contractor check splices on all branch connections of your circuits to see if they show any sign of problems.

Basic Wiring Skills

Now that you've become acquainted with wiring tools and materials, and seen how relatively inexpensive your investment will be, it's time to begin practicing the skills. But first, don't think of this as work! True, it will take some time but it certainly won't take much physical exertion. In fact, you can practice splicing and stripping insulation while sitting at a table.

Most importantly, in this chapter you'll learn how to do testing, with a voltage tester and with a continuity tester. This is an absolutely essential step in all wiring...do not neglect to read about and understand testing.

Next, you'll learn how to work with wires and cable so you can confidently make splices and connections. And you'll see that recent innovations such as push-in type terminals have made wiring even easier than before. There's also a section on connecting cable to boxes.

You'll be introduced to a service entrance panel and shown how to safely and easily work with it. At first it might look like a tangled maze of wires, but you'll soon understand what's going where. Lastly, there is a section that demonstrates fuses and the basic skills of working with them.

As serious and as potentially dangerous as wiring can be, it can also be fun. There is a certain delight in testing and discovering that you made a connection properly. But even if you don't get that little thrill, you'll find that it is a very easy skill to learn, and, once learned, it will save you a lot of money.

Using Testers to Check Your Work and to Detect Problems

The voltage tester and the continuity tester are two inexpensive, essential items for doing your own wiring (page 8). The voltage tester is primarily a safeguard against shock—used to ensure that there is zero voltage in the circuit that you are working on. The continuity tester is used to help detect electrical faults.

Using A Voltage Tester. Neon voltage testers should be of the proper voltage (about 90 to 500 volts) for use in home wiring. The voltage tester has no power source of its own; it lights when the probes are touched to anything that is charged with electricity.

The probes are designed to easily fit into the slots of a receptacle—so that you don't have to remove the cover plate. Besides being used to make sure that the power is off before making repairs, the voltage tester is also used to check if an outlet is hot, and to test for grounding.

WARNING

When using a voltage tester, be very careful to grasp the leads only by the insulated (and not the metal) areas. Also, use the probes carefully and remember that you may be near live equipment. A probe, if it accidently touches a hot and grounded object at the same time, can cause a ground fault.

TESTING

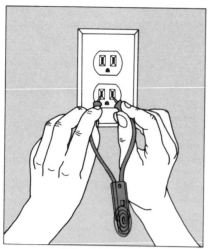

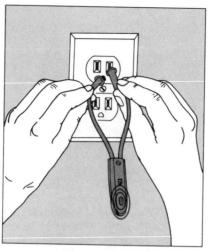

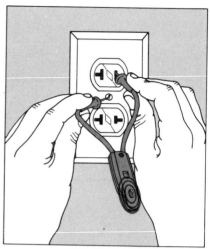

Testing a Wall Receptacle. Use a voltage tester to find out if a wall receptacle is working. Put one probe into each slot (test both bottom and top sets of slots). If the bulb glows brightly, there is power coming to that receptacle and you must shut off the power to the circuit in order to continue your work. After tripping the proper circuit breaker or removing the correct fuse, test again—until the bulb no longer glows.

Testing a Receptacle's Grounding. Use a voltage tester to check the grounding of a 3-prong receptacle. Insert one probe of the tester into the lower semicircular ground slot and the other probe into one of the elongated slots. Repeat the process but put the probe into the other elongated slot. The bulb should glow when the probe is inside the hot slot (in newer receptacles, this is shorter than the other slot). If neither slot causes the bulb to glow, the receptacle is not grounded and you should correct the wiring. If your system is not grounded, it's best to hire a qualified electrician to install a grounded system for you.

Testing for a Cover-Plate Ground. Use a voltage tester to check for grounding of the cover plate of a 2-prong receptacle—one that you have just installed or one that you intend to use with a 3-prong grounding adapter plug. Put one probe on the metal mounting screw of the cover plate (if there is paint on it, chip away some of the paint). At the same time, insert the other prong into each of the slots. The bulb should glow when the probe is inside the hot slot (in new receptacles, the shorter slot). If neither slot causes the bulb to glow, you cannot use an adapter plug and you should correct the wiring. If your system is not grounded, it's best to hire a qualified electrician to install a grounded system for you.

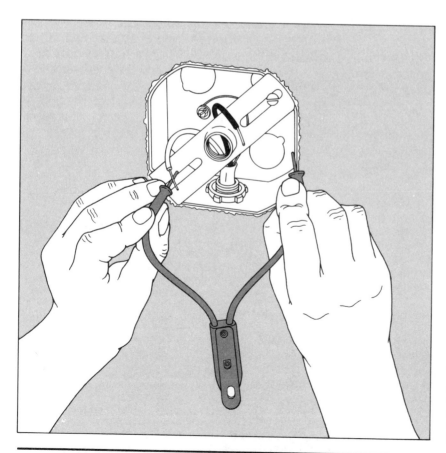

IMPORTANT

When testing for system grounding, make sure that the tester bulb glows just as brightly as it did when you tested the wall receptacle. A weakly burning bulb indicates an improper grounding path somewhere in the system.

Testing for Voltage at a Lighting Outlet. Use a voltage tester to check for voltage at a lighting outlet—to make sure that the power is off when doing repair or installation work. Turn off the power at the service panel or fuse box and set the controlling light switch to "on". (If the switch is a 3-way type, make sure that you first know exactly how to turn it on since the word 'ON' will not be printed on the toggle.) Remove the nuts or screws that hold the fixture in position and pull the fixture out of the box to expose the wires. With one hand, remove the wire caps; with the other, hold the fixture. Untwist the wires while supporting the fixture, being careful not to touch the bare copper (touch the insulation only). After setting the fixture aside, test the following with the probes: the black wire and the grounded metal box; the black wire and the white wire; the white wire and the grounded metal box. The bulb should *not* glow during any of these three tests, indicating that the power is indeed shut off.

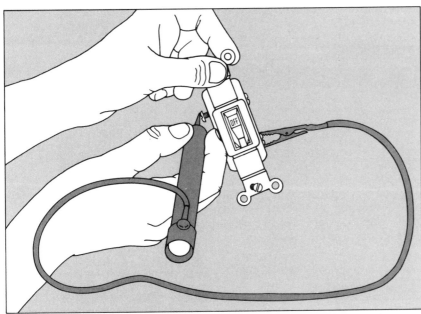

Testing a Single-Pole Light Switch.
Use a continuity tester to check for problems in a single-pole light switch. Turn off the power and completely remove the switch from its box. Clamp the alligator clip of the tester to one of the terminal screws; place the probe of the tester on another terminal screw. Flip the toggle switch off and on while holding the probe in place. The tester bulb should glow when the switch is in the "on" position and should not glow when it's in the "off" position.

Using a Continuity Tester. A continuity tester is a battery-powered device that tests for complete circuits in an appliance that is not hooked up or in a switch when the power is off. It can also be used to test a new circuit branch before hookup and to test cartridge fuses.

▊ WARNING ▊

Before using a continuity tester, make very sure that you have turned the power off, either by shutting down the circuit, pulling the fuse, or, if it's an appliance, unplugging it. *Use a voltage tester to be certain.*

Working with Conductors

The final step of wiring installation comes when you are ready to connect all of the wires you've run to each other and to the switches, light fixtures, and receptacles they'll be supplying. These connections, to be safe, must always be made within an outlet box or a junction box.

Although the skills presented here are usually done at the site of the repair or installation, you might want to familiarize yourself with them by practicing with small pieces of cable. This could save you time and errors when you take on larger projects.

▊ WARNING ▊

When testing your wiring, be sure to follow all of the safety precautions on page 5.

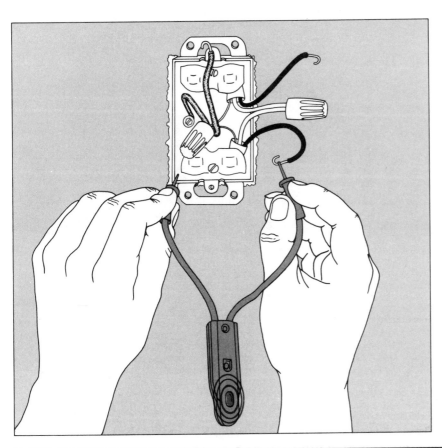

Testing for Incoming Power Supply.
Use a voltage tester to find out which black wire, if there are more than one, is carrying the current in a switch or receptacle. Turn off the power at the service panel and pull the device from the box. Disconnect the black wires from the device and bend them so that they are not touching each other or any other metal equipment. Very carefully place one probe at the end of one of the black wires and the other probe on the grounded metal box (if the box is plastic, place the probe on the bare grounding wire). Have a helper turn the power back on at the service panel. If the bulb glows, you have located the incoming power (hot) wire; if it doesn't, test the other black wire. When you have found the hot wire, tell your helper to turn off the power. Check with the voltage tester to make sure the power is off. Identify the hot wire by marking it with electrician's tape.

Stripping—Cutting Away the Insulators. In order to make connections, you must first prepare the conductors by cutting away the outer protective coverings of both cable and wires. This process is called *stripping*.

The amount of insulation that you strip from a wire depends on what you intend to do with it. If you're attaching it to a screw terminal, strip only enough bare wire to wrap ¾ of the way around the screw. If you're using a push-in type terminal, simply measure the wire with the gauge that is provided on the device.

The stripping of wires was at one time done with a pocket knife but with the invention of the handy and safer combination tool, this practice has died. Even so, you should always check to make sure that you have not nicked the wire when stripping. If you have, cut the ends and start all over.

Removing the sheathing from non-metallic cable is relatively easy and can be done in two ways: with a cable ripper or with an electrician's knife or, as an option, a pocket knife. Cable rippers generally work best when used on flat, 2-wire cable, whereas rounder, 3-wire cable is easier cut with an electrician's knife.

WARNING

When using an electrician's knife, always use a flat surface such as a wall or, if need be, a 'portable' board. Also, cut away from (not toward) your body.

Stripping cable can be tricky and might take some practice to master. The most important thing to remember is to cut gently—you can always go back and recut. But if you do make an error by nicking the insulation of an inside wire, you must cut the cable there and begin all over again.

IMPORTANT

When working on lamp switches and sockets, be sure to unplug the lamp and remove the switch and socket first (page 42).

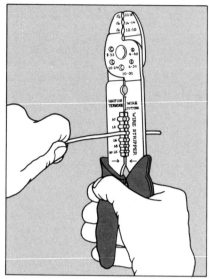

Stripping Wires. Use a multipurpose tool to strip the insulation off of wires. Determine how much wire you want to strip (for most connections this will be from ½ inch to ¾ inch). Place the wire in the proper size slot, as marked on the tool, and clamp down tightly. Next, rotate the tool back and forth until the insulation is cut. Pull the tool away from yourself and the insulation will pop off.

TESTING /CONT'D

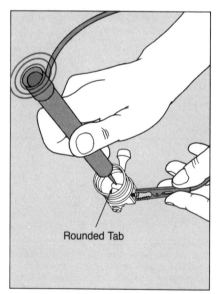

Rounded Tab

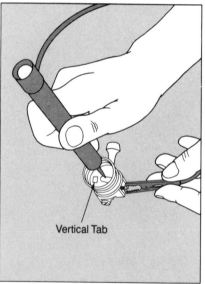

Vertical Tab

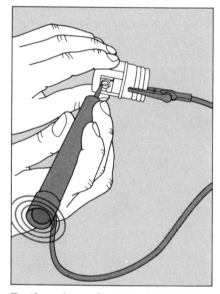

Testing a Lamp Switch. Use a continuity tester to find out if your lamp switch is faulty. First, clamp the alligator clip to the brass (hot) terminal and place the probe on the rounded contact tab in the center of the socket. Turn the switch off and on. If the testing bulb does not glow, the socket is faulty and must be replaced. If the bulb does glow but the socket had failed to light a good light bulb, then the problem might be in the contact tab—it is failing to make contact with light bulbs. To remedy this problem, simply raise the free end of the tab slightly with the tip of a screwdriver. If the lamp still does not light, the problem is not in the switch; you should next check the cord and plug (page 42).

Testing a 3-Way Lamp Switch. To test a 3-way lamp switch, use a continuity tester and begin by clamping the alligator clip to the brass (hot) terminal. Then test the four switch positions by placing the testing probe on, first, the rounded contact tab and, next, on the small vertical tab. In the first switch position the tester should light when the probe touches the vertical tab but not the rounded tab. In the second switch position the tester should light when the probe touches the rounded but not the vertical tab. In the third switch position the tester should light when the probe is at either tab and, in the fourth, or "off" position, the tester should not light at either tab.

Testing a Lamp Socket. Use a continuity tester to test a lamp socket when you're sure that the light bulb is good. Simply clamp the alligator clip to the metal screw shell and place the probe on the silver-colored (neutral) terminal screw. Both the metal screw shell and the silver-colored terminal screw should be neutral and should cause the testing light to glow. If this does not happen, then the socket has an open circuit and should be replaced.

Splicing—Connecting Wires.

Splicing simply means connecting wires to each other and because wires come in all sizes and types, there are different ways of doing this. Very small, low-voltage wires (page 9), are sometimes joined by soldering. But for most of the repair and installation work in this book, you will be splicing with what are called *solderless connectors*. Though there are slight variations in style, the most widely accepted of these are *wire nuts*.

Generally, these are cone-shaped caps of hard plastic with spirals of copper inside. The copper interior is designed to grip the wires tightly when the nut is screwed on. When purchasing, the packaging will tell you how many and what gauge wires the wire nuts can accept. When doing larger projects, you might need more than one size. In rewiring an older home you might discover that wires are connected with electrician's tape. This practice, though still acceptable, is not widely practiced today—so instructions are not presented here for the use of tape.

Wire nuts come in various sizes and colors. Check the packages to see what kind is compatible with the conductors you are using.

How a Wire Nut Works. This interior of a wire nut shows how closely the threaded sides must hug the wire in order to work properly. Also, note the amount of insulation removed from the wire. **The bare wire must be covered by the wire nut.**

STRIPPING CABLE

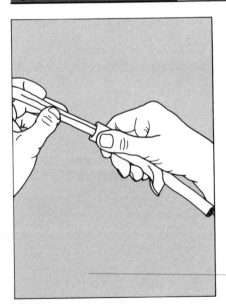

1 Using a Cable Ripper. The easiest way to strip the sheathing from nonmetallic sheathed cable is with an inexpensive cable ripper. Begin by slipping about eight inches of cable into the ripper's jaws. Next, squeeze tightly and then pull. This cuts the sheathing without damaging insulation on the conductors inside.

ELECTRICIAN'S TIP: Razor-sharp blades on knives are never a good idea. For other home repairs sharp knives are preferred, but for wiring it's actually better to use a dull knife.

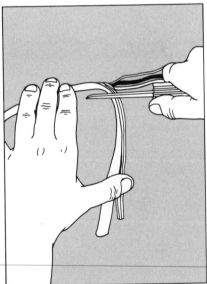

2 Cutting Off Sheathing and Filler. Peel back the sheathing and any paper or other filler. You'll find either two or three insulated wires plus a bare copper grounding wire. Cut off the sheathing and the filler with either metal shears or an electrician's knife. Always cut away from yourself and the wires, as nicking them could cause a short.

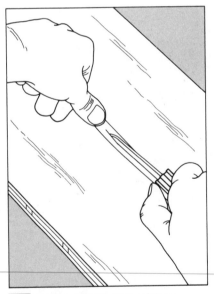

3 Optional Stripping with an Electrician's Knife. Another way to strip cable is with an electrician's knife. To do this, lay the cable on a flat surface and press your thumb close to the desired length of your cut (approximately eight inches). Now, carefully make a shallow lengthwise cut in the center of the cable. Do not press too hard or you might damage the insulation on the conductors inside. (If you do damage the insulation, begin again.) After peeling it back, cut off the sheathing and filler (step 2).

To make a splice with two solid wires, use a pair of pliers. Twist the ends a few turns until the wires are taut, snip the ends, if uneven, and then insert them into a wire nut, twisting tightly. To make a splice with a solid and a stranded wire, use a greater length of stranded wire than the solid and twist it until it becomes more compact. Then wrap it around the solid wire in a spiral pattern, slip on the wire nut and tighten it.

Use a *split-bolt connector* when you are connecting larger wires together. Also easy to use, this device has a bolt that you must tighten to make the connection. This should be wrapped with electrician's tape as an extra safety measure. (Wire nuts can also be wrapped with tape, but codes may vary as to whether or not this is necessary.)

Connecting Wires to Terminals. The other kind of connection made within boxes, besides wire to wire, is wire to terminal. This, in some ways, is the essence of wiring because it hooks up the power to the receptacles and switches that transform and modernize our homes. Connecting to terminals isn't all that difficult to learn and

yet it must be done correctly. You might want to practice it before attempting a large project.

The method of connecting is determined by the kind of switch or receptacle that you are using. Basically, there are three kinds: side-wired, back-wired, and both. Side-wired devices have binding screw terminals on the side; back-wired devices have slots on the back; the others are a combination of the two.

A little more manual dexterity is needed to use the screw terminals because you must wrap the wire around the screw, whereas with the push-in type you merely insert the correct amount of bare wire. Consequently, the push-in types are a bit more expensive, but the ease of use might be worth it.

One rule that is critical in making connections to terminals: *Never connect more than one wire to a terminal.* You may connect two or three wires to *each other* but never to a terminal. If you have a wiring situation where you need to connect two wires to a terminal, then you must use a *pigtail,* page 31, a short 'jumper wire' that will connect the two wires plus a wire from the terminal in a wire nut.

Connecting Nonmetallic Cable to Boxes

Metal boxes, which are covered on page 75, have removable disks built into them which are called *knockouts.* These provide the openings by which cable enters the box. You may remove a knockout before installing a box into a wall, or you may remove a knockout to a previously installed box for another connection if it is large enough (page 74). In any case, *you may not open a knockout and then not utilize it.* You must fill in the hole with a special knockout seal if you make such an error.

SPLICING...CONNECTING WIRES TO EACH OTHER

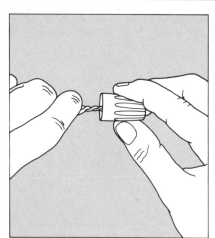

Connecting Solid Wires with a Wire Nut. To splice two solid wires, cut off about ¾ inch of insulation from each wire. Hold the wires side by side and twist them clockwise around each other with a pair of pliers, tautly. If the ends are not even, snip off the longer wire. Insert them into the wire nut and screw the nut as tightly as possible. Be sure that no bare conductors are exposed; if they are, remove the wires, snip away a small amount of wire and reinsert, again screwing the nut as tightly as possible.

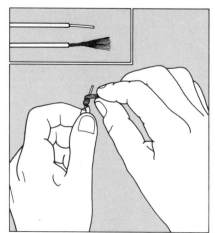

Connecting Solid and Stranded Wires with a Wire Nut. To splice a solid and a stranded wire together, first cut off an extra ½ inch of insulation from the stranded wire. The solid wire will measure about ¾ inch and the stranded wire about 1¼ inch. Next, twist the thin stranded wires around each other to create a more compact 'wire'. Wrap this around the solid wire in a spiral pattern. Insert them into the wire nut and screw the nut as tightly as possible. Be sure that no bare conductors are exposed; if they are, remove the wires, snip away a small amount of wire and reinsert, again screwing the nut as tightly as possible.

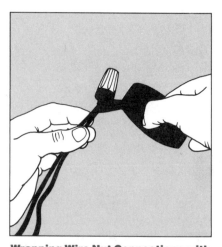

Wrapping Wire Nut Connections with Tape. Many electricians secure their connections by wrapping electrician's tape around the wire nut and the wires. This ensures that the connection will not jar loose as it is pushed into the box. Simply wrap the tape around the base of the wire nut, a few times around the wires, and then once more around the wire nut.

ELECTRICIAN'S TIP: Left-handed persons: Be sure to twist wires clockwise and the wire nut in the same direction.

USING PUSH-IN TYPE TERMINALS

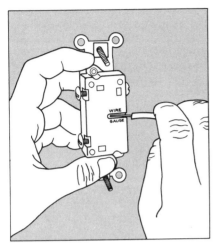

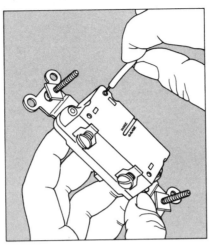

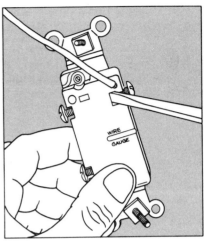

1 **Measuring the Wire.** Often, receptacles and switches have screws on the sides and also have features on the back that allow you to connect the wire by inserting it. These push-in type terminals are very easy to use, provided that you follow specifications. On the back of the switch or receptacle is a wire gauge that shows how much insulation you should strip from the wire. This will generally be about ½ to ¾-inch, but be sure to measure with the gauge.

2 **Inserting the Wire.** Insert the bare wire into the round terminal aperture and push until a metal spring grips it.

3 **Disconnecting the Wire.** If you need to release the wire for any reason, simply insert the tip of a small screwdriver or a piece of stiff wire into the release aperture and pull out the wire.

USING SCREW TERMINALS

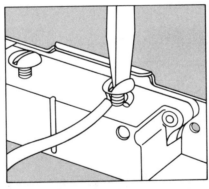

1 **Connecting the wire.** Strip only enough insulation from the wire to allow the bare end to be wrapped ¾ of the way around the screw. With needle-nose pliers, bend the bare end into a loop. Loosen the terminal screw and hook the screw so that when it is tightened it will help close the loop.

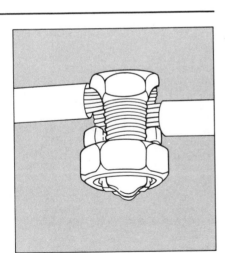

Using a Split-Bolt Connector. These easy-to-use connectors are normally used for large wires. Make sure you are using a connector made of metal that is compatible with your wires. Strip the wire ends, place them in the connector, and tighten the bolt. Then wrap the connection with electrician's tape to the same thickness as the wire insulation.

Wrap Wire Clockwise

Tighten Screw Clockwise

Full Contact

2 **Wrapping the wire correctly.** When connecting wire to a screw terminal it is very important that the wire be wrapped in the same direction that the screw will be tightened (clockwise). Also, the screw must be tightened to full contact.

WARNING

Never connect more than one wire to a screw terminal. If you need to use one screw with two wires, you must use a pigtail and a wire nut, page 31.

REMOVING KNOCKOUTS

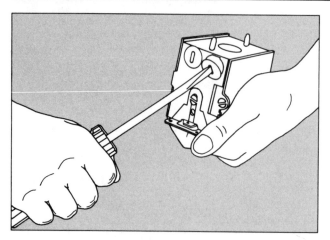

Removing a U-Shaped Knockout. To remove a U-shaped knockout, insert a screwdriver into the slot, pry the piece away from the box and work it back and forth until it is free. (If you are removing a knockout from a previously installed box, begin from within the box with a hammer and nail set.)

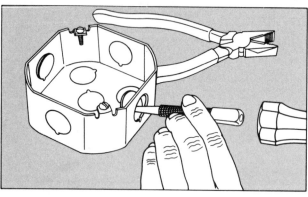

Removing a Round Knockout Place the box on a flat surface and lightly tap the knockout with a nail set and hammer. Grasp it with a pair of pliers and move it back and forth until it is free.

There are two kinds of knockouts: U-shaped and round. The U-shaped knockouts have small slots for removing with a screwdriver and are found almost exclusively in wall boxes. The round knockouts have no slots, must be removed by hammer and nail set, and are found in ceiling, junction, and wall boxes.

Once you have determined how to get cable into a metal box, next you must decide how to secure it. Although there are other types of clamps available, the two most common types used in home wiring are the *internal clamp* and the *two-part* connector.

The internal or 'saddle' clamp is a built-in feature of some boxes. It can be used with nonmetallic cable and there is also a type used only with armored cable. Internal clamps accept cable entering U-shaped knockouts.

WARNING

When removing knockouts from metal boxes, for safety wear eye protection.

Another clamp often used with metal boxes must be purchased separately, the two-part or 'locknut' connector. It consists of a clamp that holds the cable and is connected to a threaded tube. The

CONNECTING CABLE IN BOXES

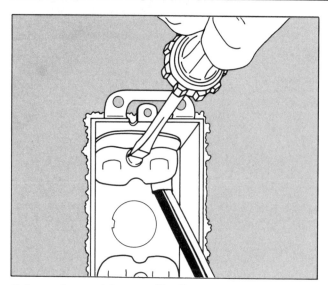

Using an Internal Clamp. First, if you are using nonmetallic sheathed cable, make sure that the clamp looks like the one shown here. Now loosely screw the clamp into place, prepare your cable and wires by cutting the sheathing and stripping, and pull the cable through the knockout and under the clamp. Make sure that the clamp is contacting the cable sheathing (not the conductors), and screw it tightly into place. The sheath of the cable must extend *no less* then ¼ inch into the box.

Using a Nonmetallic Box. Most nonmetallic boxes have perforated openings; to use them simply pry them open with a screwdriver and insert the cable. These don't anchor the cable as do clamps in metal boxes. By Code, nonmetallic cable may be unclamped in single nonmetallic boxes—as long as the cable is secured with a staple or strap within 8 inches of the box. The cable sheath must, however, extend through the box opening by at least ¼-inch.

Basic Wiring Skills**

tube fits into the knockout hole and is secured with a locknut. This type is always used with round knockouts.

Connecting Cable to Nonmetallic Boxes. Nonmetallic boxes, which are gaining in popularity, offer an advantage in connecting cable. They have perforated tabs which are easily pried open for the insertion of cable. The cable is not truly clamped to the box; this procedure is permitted by Code with a few minor exceptions to the rule (below, left).

Installing Circuits at the Service Entrance Panel

When adding circuits in your home, the last step will be to connect the cable to the service entrance panel. This is done by connecting the wires to the neutral bus bar and also to circuit breakers.

Codes differ in regard to this work, but generally you are permitted to do it yourself. However, you may need to have it inspected after it is completed.

Once you have determined that the power is off, you can begin to hook up your circuit. If you are connecting more than one circuit, do them one at a time to keep the wires straight. Note: The service entrance panel is also discussed on pages 87-89. You might want to review this material before proceeding.

WARNING
Whenever working at the service entrance panel, be absolutely sure that you have switched the main disconnect switch to OFF. This main switch is usually located in the service panel, but if it is not you should seek the assistance of a qualified electrician, an electrical inspector, or your utility company. Also, see the safety precautions on page 4.

Pull each cable into the panel and identify it. Mark it on a chart on the cabinet door. For each cable connection, strip the sheathing from the cable and then remove about ½ inch of insulation from the wires.

Here are the basic steps for hooking up the three kinds of home circuits: a 120-volt, a 120/240-volt, and a 240-volt circuit. Be sure to refer to the accompanying illustrations as you follow these steps.

Connecting 120-Volt Circuits.

1 Connect the white wire and the the grounding wire to the neutral bus bar as shown.

2 Connect the black wire to the circuit breaker as shown.

3 Insert the circuit breaker into the panel as shown.

Connecting 120/240-Volt Circuits.

1 Extend the white neutral wire around the edge of the cabinet to the neutral bus bar and insert it under the screw terminal. Tighten the screw.

2 Insert the black hot wire under a terminal on a double-pole circuit breaker and tighten the screw.

3 Insert the red hot wire under the other terminal on the double-pole circuit breaker and tighten the screw.

4 Extend the grounding wire to the neutral bus bar, insert it under a terminal and tighten the screw.

Connecting 240-Volt Circuits.

1 Connect the black wire to a terminal on a double-pole circuit breaker.

2 Connect the white wire to the other terminal on the circuit breaker. Either paint it black or wrap it with electrician's tape to identify it as a hot wire.

3 Connect the grounding wire to the neutral bus bar.

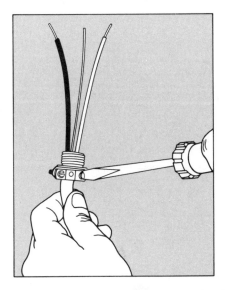
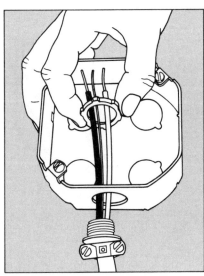
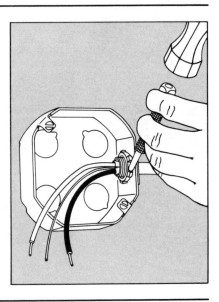

USING A TWO-PART CONNECTOR

1 Fastening the connector. Prepare your cable and wires by cutting the sheathing and stripping insulation. Slip the connector over the cable so that it is contacting the cable sheathing and the threaded end faces the stripped wires. Tighten the clamp screws onto the cable.

2 Assembling the connector. Slip the wires and the threaded end of the connector into the box through the knockout hole. Slip the connector locknut over the wires and screw it onto the threaded bushing.

3 Fastening the connector to the box. Attach the box to the wall or ceiling and tighten the locknut using a hammer and nail set.

CONNECTING CABLE AT THE SERVICE ENTRANCE PANEL

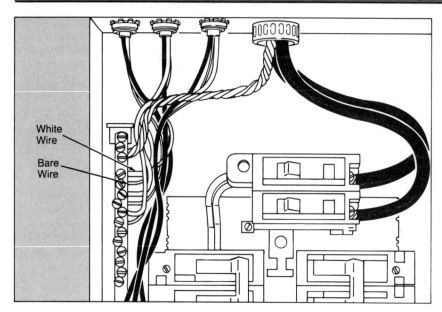

White Wire

Bare Wire

Wiring at the Neutral/Ground Bus Bar.
When wiring a 120 or a 120/240-volt circuit, you must always connect both the white neutral wire and the bare copper grounding wire to the neutral/ground bus bar. Extend the wires around the edge of the cabinet to the neutral bus bar, cut off any excess length, strip approximately ½ inch of the neutral wire, and insert it into the terminal. Tighten the screw. Next, insert the bare grounding wire into the terminal and tighten the screw. Keep an orderly alternating arrangement of these sets of wires to make it easier when troubleshooting.

ELECTRICIAN'S TIP: Never connect the bare grounding wires and white neutral wires together in a panel that is *not* part of the main service equipment. Neutral and grounding wires must be isolated from each other in other subpanels; these subpanels also have special rules regarding installation of their bus bars.

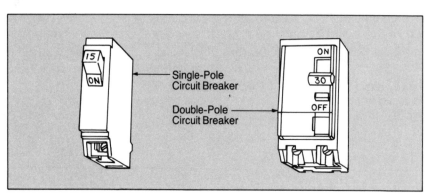

Single-Pole Circuit Breaker

Double-Pole Circuit Breaker

Selecting Circuit Breakers.
Choose the type of circuit breaker you need according to the amperage of the circuit that it protects. The capacity, rated in amperes, is marked on the levers. A *single-pole circuit breaker* is used for all 120-volt circuits. A *double-pole circuit breaker* is used for all 120/240-volt and 240-volt circuits.

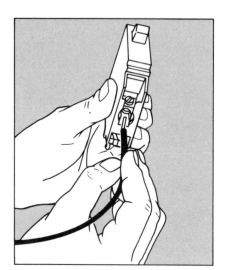

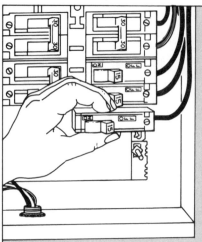

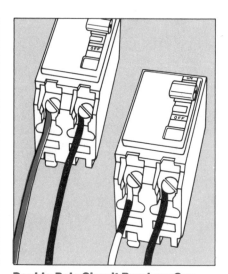

Inserting Wire into the Circuit Breaker Terminal. Shown is a single-pole terminal and a black hot wire. With the main power switch off, strip approximately ½ inch of the wire and insert it into the terminal under the screw. Tighten the screw. Use the same process when using double-pole circuit breakers; insert a red wire (for a 120/240-volt circuit) or a white wire permanently reidentified as black by tape or paint (for a 240-volt circuit).

Snapping the Circuit Breaker into the Panel. With the main power switch off, hold the breaker at an angle and hook the end nearest the wire first. Push the other end down until the contact point locks over the bus bar tab.

Double-Pole Circuit Breakers Connected to 120/240-Volt and 240-Volt Circuits. Shown are connected double-pole circuit breakers. Note that the 120/240-volt circuit provides two hot wires, black and red. The 240-volt circuit also provides two hot wires, but one of them is white which normally indicates neutral. This white wire must be marked black, with either electrician's tape or paint, to permanently reidentify it as a hot wire.

Working with Fuses

Some homes still have fuse panels at the service entrance instead of the easier to use circuit breaker panels. If yours is such a home, you should have a basic understanding of fuses and learn how to work with them when problems arise.

Fuses are, very simply, thin pieces of metal that are designed to melt quickly when too much current flows through them. This is commonly called a 'blown fuse'; as a result the current is stopped and the circuit shut down.

There are several kinds of fuses, although most people are familiar with the standard *plug* (screw-in) type. Two variations of the plug type are the *time-delay* and *type S* fuses. These have additional features that make them safer or easier to use with certain motor-operated appliances. These fuses are usually encased in glass so that you can see their amperage rating and you can also see their metal strips.

Cartridge fuses are not made of glass; they come in two different types: ferrule-type and knife-blade. These fuses should be checked with a continuity tester to determine if they are blown since they give no visible evidence like the glass-encased types.

Why Fuses Blow

Fuses blow for one of three reasons:

1 If a circuit is overloaded. If too many appliances are demanding more electricity than the circuit can provide, a fuse will be blown. A glass-encased fuse will look clear because the overload gradually heated the metal strip of the fuse.

2 A short circuit. If there is a problem in an appliance or in appliance wiring, such as two wires touching, a fuse will be blown. A glass-encased fuse will look cloudy or blackened because the fuse was blown immediately.

3 A loose fuse. Make sure that the fuse is screwed in properly. Though rare, this can be a cause of blown fuses.

By looking at a plug fuse, you can guess what blew the fuse. Further clues are: If a fuse blows immediately after plugging an appliance in, no matter where you plug it in, the trouble is most likely in the appliance. If, however, a fuse blows routinely in a particular circuit, you probably have an overload and you either need to recalculate the load to eliminate appliances (page 51) or consider upgrading your wiring.

WARNING

When changing fuses always check the fuse you are replacing for amperage rating. If you use a larger amperage fuse than the circuit can hold, you could cause the wiring to overheat and possibly start a fire. For example, if a 15-amp fuse continually blows, don't consider replacing it with a 20-amp fuse. This is a very dangerous practice. Also, when buying a home that has a fuse panel, it's wise to check all of the fuses for amperage rating.

ELECTRICIAN'S TIP: Plug fuses of 15 amp rating can be identified by a hexagonal configuration in the window, cap or elsewhere…to help distinguish them from higher-amperage fuses.

A TYPICAL FUSE PANEL

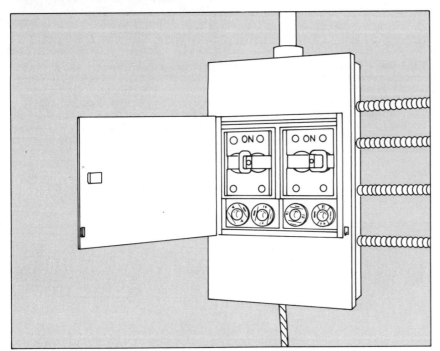

This typical fuse panel contains two types of fuses. The main fuses are located in the pull-out blocks at the top of the panel. Each block contains two ferrule-type 35-ampere cartridge fuses. Below these are plug fuses which protect the separate branch circuits. Some of these are 15-ampere fuses, such as those protecting general lighting circuits. Others are 20-ampere fuses which are used for circuits with appliances such as refrigerators, irons, and toasters.

WARNING

While working at the fuse panel, follow the safety precautions on pages 4-5.

To remove any fuse, either to work on the circuit or to replace it, first turn off all the lights and appliances controlled by that fuse. Knife-blade and ferrule-type cartridge fuses are removed by pulling out the correct block and continuing with a fuse puller (page 8). Remove other types of fuses by touching only the insulated outer rim while unscrewing.

TYPES OF FUSES

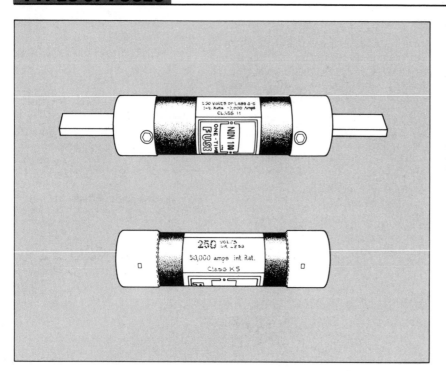

Cartridge Fuses. There are two types of cartridge fuses used in homes. The *knife-blade* fuse has metal blades that snap into spring clips. It is designed for more than 60 amps and is commonly used to protect the entire service entrance. The *ferrule-type* fuse has metal caps at each end that snap into spring clips. It is rated from 10 to 60 amps and is often used to protect circuits for large individual appliances. Because of their design, blown cartridge fuses show no sign of damage; therefore, they must be tested with a continuity tester (shown at right).

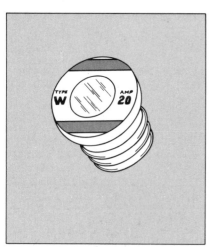

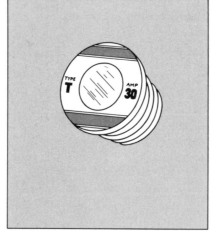

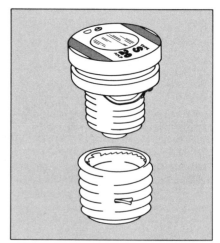

Plug Fuse. Also called an *Edison-base* fuse, this, the most common type, screws into the service panel. It is rated up to 30 amps, with the rating visible through the protective glass cap. To determine why a plug fuse is blown, inspect the top of the fuse. If the glass is blackened, suspect a short circuit. If the metal strip is melted but the glass is clear, the problem is an overloaded circuit.

Time-Delay Fuse. Similar to the screw-in type plug fuse, the time-delay fuse has a specific feature for use with certain appliances. The time-delay fuse will blow immediately when there is a short circuit. However, it will very gradually melt when there is a slight overload. These are very useful if you are using equipment or tools with motors that require extra power only when they are first turned on. The fuse is designed to allow for a time delay before it blows.

Nontamperable Fuse. Also known as a *Type-S* fuse, this contains a safety feature that prohibits you from using a higher amperage rated fuse than you should for a given circuit. These fuses come with adapters that screw and lock into the fuse socket. The fuses are sized by their amperage rating—so once you've installed a 15-amp adapter you cannot put a 20-amp Type-S fuse into it because it will not fit.

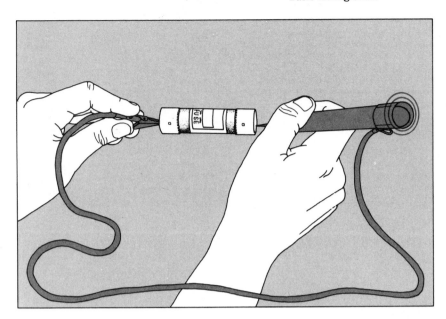

Testing a Cartridge Fuse. When an appliance that is protected by a cartridge fuse fails to operate, you cannot tell if the problem is the appliance or the fuse—because cartridge fuses give no external evidence of being blown. The most efficient way of finding the problem is to check the fuse with a continuity tester (page 8). Remove the fuse and touch one metal end with the clamp and the other with the probe. If the bulb glows, the fuse is good and the problem is in the appliance. If the bulb does not glow, the fuse is blown; replace it with a new one. If the appliance still doesn't work, it should be checked by a repairman.

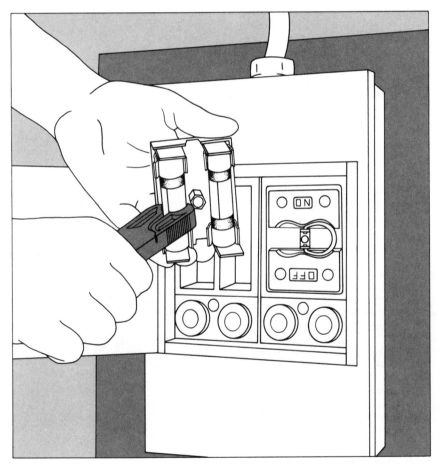

Replacing a Cartridge Fuse. To replace a cartridge fuse in a service panel, as shown here, begin by pulling out the panel block. With a plastic fuse puller (page 8), firmly grasp the middle of the fuse and pull hard to release it from the grip of the spring clips. Do not touch the end caps of the fuse—they might be hot. Insert the replacement fuse by hand.

Cartridge fuses are sometimes used in auxiliary fuse boxes, such as those installed to protect a clothes dryer circuit. To remove fuses in these boxes, open the box by moving the cutoff lever to OFF; this will expose the fuses. Then proceed with the above steps.

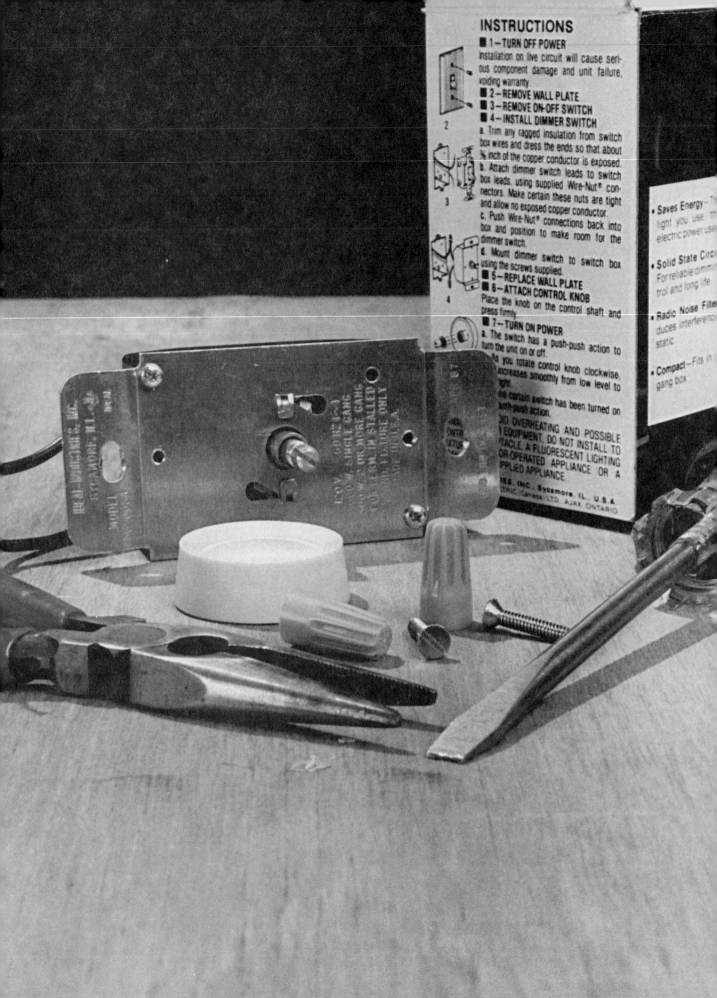

Common Wiring Repairs and Improvements

Now that you fully understand the importance of safety and testing and have practiced the basic wiring skills, it's time to look around your house with a critical eye. There might be a defective switch or receptacle that you've been overlooking...now is the time to replace it!

More often than not, however, switches and receptacles aren't broken—after all, they are very sturdy devices. Instead, they might be outdated or simply not performing all of the various functions that they could perform. You might have an older home with switches that make a loud snap when flipped on. All standard switches are now the quiet type and are installed very easily.

Chances are that even if you have a relatively new home, your home was wired in a 'plain vanilla' style—with standard switches and receptacles. With today's array of specialized switches and receptacles you can transform your home's electrical system to suit your own specific needs. Kitchens, workshops, garages, and children's rooms are just a few areas where you can make electricity work for you in new ways.

In this chapter, you'll learn how to make many improvements for a safer, more convenient, and energy-conserving home. You'll also gain plenty of experience testing wiring and working at outlets; so when the time comes to install new wiring, you'll be well prepared!

Working with Wall Switches

Normally, wall switches are associated with lighting, although they can be used to give power to a receptacle or to turn on appliances such as garbage disposals. In any case, they last a long time, so once you've installed quality switches, properly, you won't need to make repairs often.

Four Basic Types. Switches come in four basic types which you must familiarize yourself with if you are doing any replacements or installations.

They are: *single-pole, three-way, four-way,* and *double-pole.* The names refer to the number of terminals on the switches. When replacing a switch, be sure to replace it with the same basic type as was formerly in the box.

Within these four basic types, there are also different kinds of switches, named according to how they are wired —*side-wired* and *back-wired/side-wired.* When purchasing switches, ease of use and expense will be top considerations. Although back-wired/side-wired switches are more expensive, you might

consider them for their quick push-in terminals.

As always, another consideration when purchasing switches is safety and that's why you should look for the Underwriters' Laboratories marking. It and other important information is printed or stamped directly onto switches. Although it is not required by Code, most new switches come with a green-colored grounding terminal, an additional safety feature. Purchase this type if it is available.

Mood-Setters, Helpers, and Energy-Savers. Wall switches simply aren't what they used to be. No longer does the power have to be immediately switched on or off. Now there are specialized switches for special areas of the home such as workshops and guest rooms. And there are switches for every room that offer more safety and convenience than the regular kind.

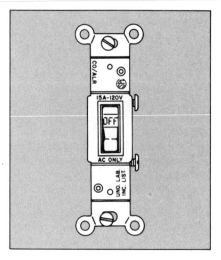

Reading Information on a Switch. On the front of a switch you will find information that verifies it to be acceptable for use. The tester's marking, usually UND.LAB.INC. LIST., certifies that it has met the Underwriters' Laboratories, Inc. standardized tests. Voltage and amperage ratings assure you that you're using the correct switch. This switch can handle up to 15 amps of current at a maximum of 120 volts. The type of current can be shown in two ways, either as AC ONLY or combined with the amperage/voltage (example 15A-120VAC). AC stands for alternating current and is the almost universally accepted designation that you should look for. CO/ALR indicates that you can use either copper or aluminum wiring with this switch.

FOUR TYPES OF SWITCHES

Learn to tell switches apart by looking at the number and color of terminal screws, and by looking at the switch toggle for ON/OFF markings.

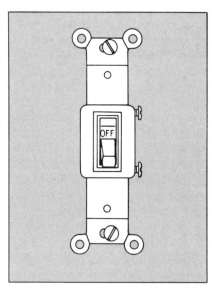

Single-Pole Switch. Commonly used, this switch controls a light or receptacle from only one location. Note the two brass-colored terminals and the ON/OFF markings on the toggle.

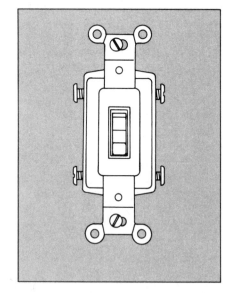

Four-Way Switch. Used in conjunction with three-way switches, these switches are used to control a light or receptacle from three or more locations. Note the four brass-colored terminals and the absence of ON/OFF markings on the toggle.

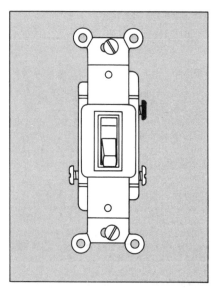

Three-Way Switch. Used in pairs, these switches control a light or receptacle from two locations. Note the terminals—two brass-colored and one black or copper-colored. Also, there are no ON/OFF markings on the toggle.

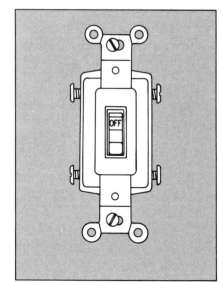

Double-Pole Switch. Double-pole switches are used to control 240-volt appliances. Like four-way switches, they have four brass-colored terminals but note the ON/OFF markings on the toggle.

POSITIONS OF TERMINALS

Once you have determined the function of your switch you will also have a choice of switches according to the positions of their terminals. Make your choice according to how it will be arranged in the box and its ease and convenience of use.

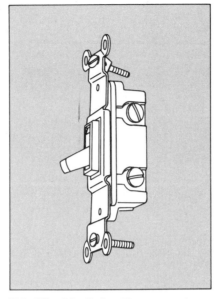

Side-Wired Switch. Most commonly used. Terminals are located on the sides; cable wires will be wrapped around and screwed to terminals.

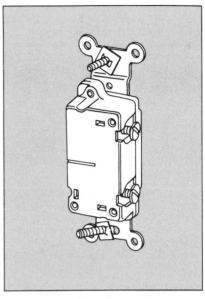

Back-Wired/Side-Wired Switch. Switch contains terminals of two types; cable is connected by either side-wired screw terminal or back-wired push-in terminal.

SPECIAL PURPOSE SWITCHES

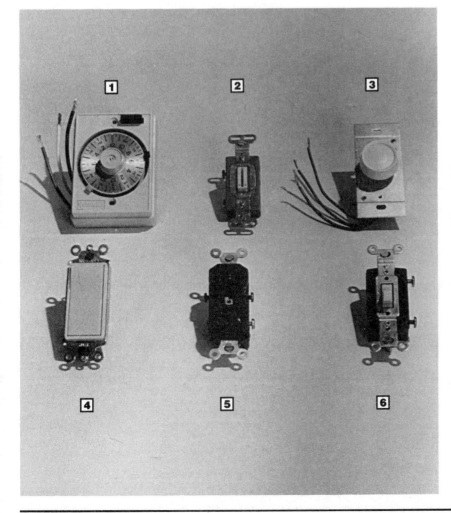

1 Time Clock Switch. Turns lights or circuits off and on at preset times. Used as a security device.

2 Locking Switch. Tamper-proof, comes with a key that turns the power on. Good for workshops and other potentially dangerous areas.

3 Dimmer Switch. Controls the lighting level to suit your needs and moods. Shown is a rotary style switch; a more expensive touch type is available.

4 Rocker Switch. Very decorative but more expensive than a regular switch, also quiet.

5 Pilot Light Switch. The pilot light lets you know whether or not you switched power off in an area that you can't see, such as a garage or basement.

6 Lighted-Handle Switch. The light in the handle serves as a 'night light'; makes it easier to find the switch. An appropriate switch for guest rooms.

REMOVING AND REPLACING A SINGLE-POLE SWITCH

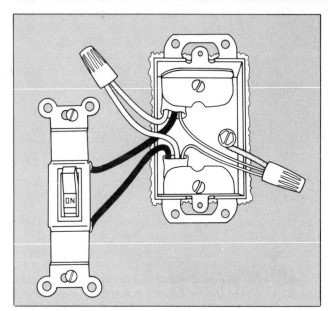

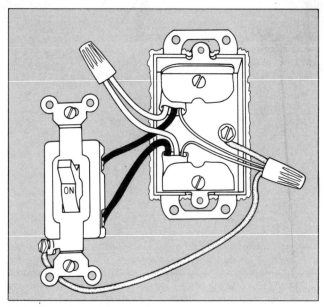

Removing an Old Middle-of-the-Run Switch. Shut off the power to the circuit at the service panel. Remove the wall plate and test with a voltage tester to determine if the power is off. Remove the screws from the yoke and pull the switch out of the box. Note that this old-style switch has no grounding terminal screw (old switches were grounded by way of the mounting screws which gave them contact with the grounded boxes). The white wires are connected by a wire nut; do not disturb them. Loosen the terminal screws holding the black wires or, if it's a push-in type, remove the wires (page 19).

Installing a New Middle-of-the-Run Switch. Insert the black wires under the terminal screws of the new switch and tighten the screws down (page 19). With a single-pole switch it doesn't matter which black wire goes to which terminal. With a push-in type make sure that the wires are not inserted into holes marked 'white'. Your new switch should have a green-colored screw on it; this is the grounding wire terminal. Take a small piece of bare copper or green coated wire (the same gauge as your cable wires) and connect it to this terminal. Next, connect the other end of this wire to the wire nut where the bare copper wires from the cables and the green grounding wire in the box are already joined. Gently push the switch back into the box and fasten it with the mounting screws.

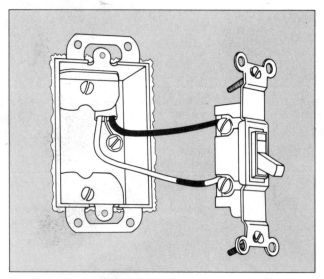

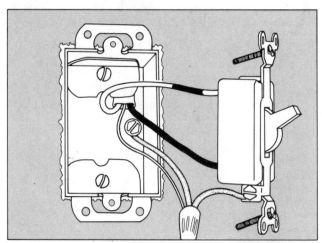

Removing an Old 'Switch Loop' Switch. Follow the above steps to get the switch out of the wall. When you pull an old switch from a wall and see only one cable or two wires coming into the box, you have a switch loop. The black wire should be on the upper terminal; the white one on the bottom. The white wire should also be marked black, either with electrician's tape or paint, as it is also a hot wire. As was the case with an old middle-of-the-run switch, there is no grounding wire terminal on this switch.

Installing a New 'Switch Loop' Switch. Connect the black and black-marked white wire to the brass-colored terminals of the single-pole switch. Then connect a short piece of bare or green wire from the grounding terminal of the new switch to the wire nut that is holding the box and cable grounding wires. Gently push the switch back into the box and fasten it with the mounting screws.

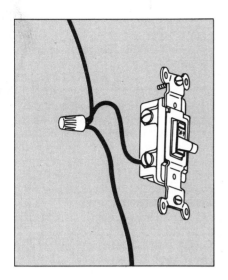

A Pigtail. When it is necessary to connect several grounding wires to a single terminal, you cannot because of this rule: *Never connect more than one wire to a terminal.* Whenever this occurs, you must use a pigtail. A pigtail is merely three or more wires spliced together and joined by a wire nut. A short wire from this wire nut is connected at the switch (or other) terminal. When counting wires in a box to determine box size, count each pigtail as one.

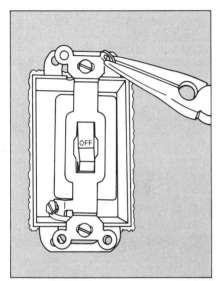

Mounting New Switches into the Wall. When installing any new switches, make sure that they are vertical. For switches with no ON/OFF markings the switch should be on when the toggle is up and off when it is down. Boxes are often crooked but mounting screw slots are wide enough for you to adjust the switch. Each end of the yoke has a pair of ears that will keep the switch flush with the wall if the box is recessed. If the box is already flush and these ears are in the way, snap them off with pliers.

Dimmer switches, almost a mainstay in modern homes, give you not only the benefit of mood-setting but the pleasure of a lower energy bill. Other switches contain timers that make your life easier and more secure, and still others add stylish decorative touches to your home. Before making a replacement with a regular switch, note the location, what you're lighting or powering, and then consider a special purpose switch.

Wiring Runs—An Explanation. When replacing or installing switches, you should determine where the box is positioned in relation to the circuit; this position within the wiring run determines how many wires there are in the box and how they should be connected to the switch. There are only two possibilities: *middle-of-the-run* and *end-of-the-run* wiring.

In a middle-of-the-run box you will find two cables, one coming in and one going out, indicating that your switch is in the middle of the wiring run. In an end-of-the-run box you will see only one cable with its set of wires, signifying that this switch is the last outlet on the circuit. Another more specific name for this is a *switch loop.* In a switch loop both wires are hot, carrying power only to and from the switch.

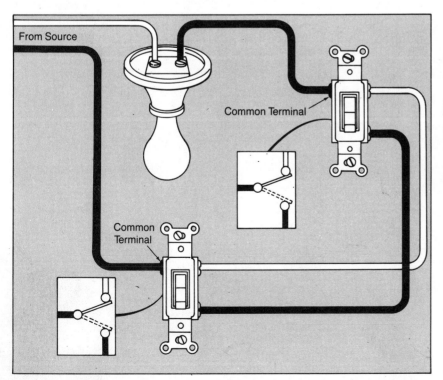

From Source

Common Terminal

Common Terminal

A Three-Way Switch Circuit. Three-way switches, useful in hallways, on stairways or in any room that is entered and exited often and from different doors, make it possible to operate lights from two different switches. The toggles on a pair of switches are not marked ON or OFF because an up or down position can turn the light off or on depending on the position of the toggle on the other switch. Each switch contains three terminals, hence the term 'three-way'. One of the three is darker colored; it is called the *common terminal* because the hot wire coming from the service panel is attached to it. The other two lighter colored terminals are called *traveler terminals.*

Not all three-way switch wiring runs will be exactly like this one; the light could be in a different position in relation to the switches or the power source could enter the circuit differently. But this diagram demonstrates how the switches, with their three terminals, work. With one switch up and one down, the circuit is open and the light is off. By flipping either switch so both are up or down, the circuit is completed and the light goes on.

A Faulty Switch—Start to Finish. Here are the basic steps that you should take to determine whether or not you have a faulty switch, and, if so, to replace it with a new good one. Be sure to follow all of the aforementioned safety precautions and refer to the specific illustrations and directions for the kind of switch you are installing.

1 If a light fails to come on when you flip the switch, first check the light bulb.

2 Check the circuit breaker or fuse to make sure that the power is on.

3 Turn off the power.

4 Test with a voltage tester to make sure that the power is off. Touch one tester probe to the side of the metal box and the other probe to each switch terminal. The tester should not glow in any position.

5 Remove the switch and test with a continuity tester to see if it is faulty (page 15).

6 Connect the wiring of the new switch and install it properly (page 30).

7 Test by turning on the power.

Three-Way and Four-Way Switches. The terms 'three-way' and 'four-way' are somewhat misleading but as mentioned earlier (page 28), they indicate the number of terminals, not the number of switches in use. As shown in the diagram, a light controlled by three-way switches is controlled from two and not three switches. Likewise, four-way switches operate from three positions and not four.

If you have a problem in a three-way circuit, you should test both switches to see which is faulty. When testing a four-way circuit, begin by testing the three-way switches. In the replacement process, be sure to follow safety precautions because all of the wires, regardless of color (black, red or white) are hot. Also, pay close attention to how the old switch was wired since you will have to duplicate it.

Installing Special Purpose Switches. Before purchasing special purpose switches, make sure that you know where, within the circuit, your switch is. Not all of these switches can be installed in end-of-the-run wiring or in three-way/four-way circuits. However, installation is relatively easy, since wire colors are standardized, plus some devices have lead wires which save you the trouble of making connections to terminals.

Dimmer switches come in various styles although the control knob type is the most popular; it allows you to brighten and dim a light within its full range. When purchasing dimmer switches, remember to calculate the total wattage of the fixture that the switch will control. The watts rating on the front of the switch must be lower than or equal to this figure.

Occasionally, dimmer switches will cause interference with television sets or AM radios. If this happens in your home, you can likely remedy the problem by either moving the appliance to another outlet or circuit, or you can purchase a power-line filter. These small devices, available at radio supply stores, fit between the cord and the receptacle to help trap house wiring interference.

TESTING A THREE-WAY SWITCH

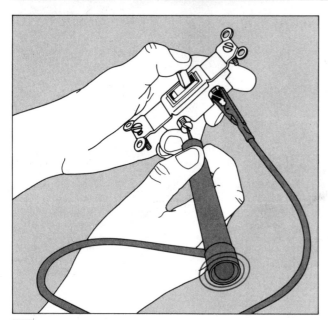

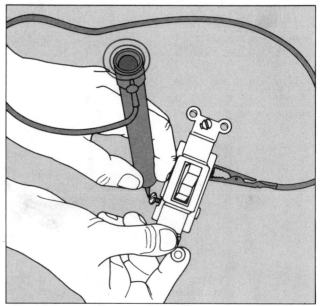

1 When a three-way switch circuit is faulty you need to test both switches with a continuity tester to find out where the problem is. Begin by turning the power off. Disconnect one of the switches (upper right), and attach the continuity tester to the common terminal by its clamp. Place the tester's probe on one of the traveler terminals and move the switch toggle up and down. The tester bulb should glow when the toggle is either up or down.

2 Next, keeping the toggle in the position that caused the bulb to glow, place the probe on the other traveler terminal. The testing bulb should not glow now, but it should work when the toggle is flipped to the other position. If the switch has passed both of these tests, it is good; reconnect it and then conduct the same tests on the other switch.

REPLACING SWITCHES

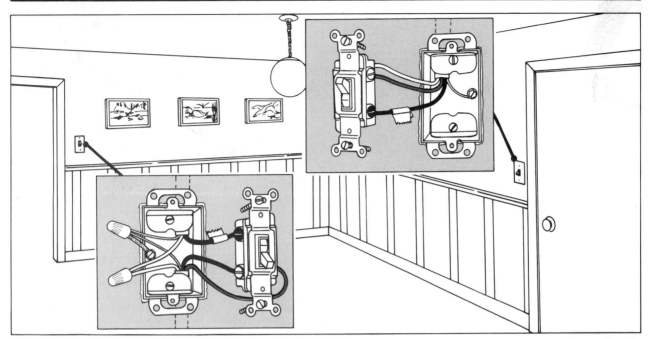

Replacing Three-Way Switches.
Shown above is a typical room that contains a pair of three-way switches and the switches themselves after they are correctly installed. After determining which of the pair is defective by testing, begin the replacement procedure as follows: Remove the defective switch from the box. Turn the power off and test to make sure that it is off.

Next, mark the wire connected to the

common terminal with a piece of masking tape. (The common terminal might be marked COMMON or it will be of a darker color than the other terminals, either black or copper-colored.)

Remove all of the wires and reconnect them as they were previously connected to the old switch. A possible exception is the grounding terminal on the new switch. You might need to make a pigtail to the ground

wires (page 31) and possibly a pigtail to the box terminal.

The red and white (or white marked black) wires can be connected to either of the traveler terminals. Most importantly, however, the taped black wire must be connected to the common terminal. If the switches do not turn on the lights from two locations, reverse the travellers.

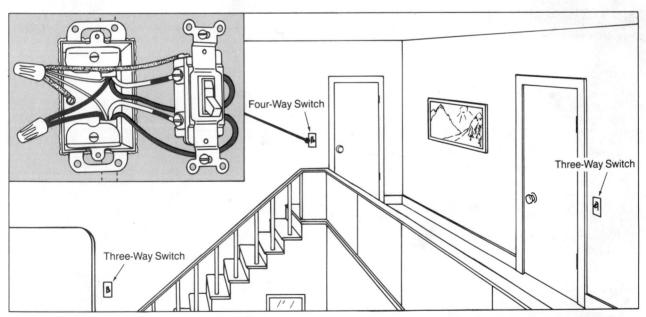

Four-Way Switch

Three-Way Switch

Three-Way Switch

Replacing a Four-Way Switch.
Four-way switches, so-named because of their four terminals, are used along with three-way switches to provide additional points from where a light can be switched on and off. One or more four-way switches can be used, but they must always be positioned in the circuitry between the three-way switches.

Before replacing a four-way switch, turn the power off to the circuit you are working on

and test to make sure that it is off.

All four wires connected to the old switch terminals will be hot wires though they may be red or white wires marked black or black wires. In any case, there will also be two black wires in the box connected by a wire nut; do not touch these wires. Beginning with the top wires of the old switch, remove them and connect them to the top terminals of the new switch. Next, repeat this process with the bottom set of wires.

Gently push the switch back into the box, reattach the wall plate and turn the power back on. Test the new switch and if it doesn't work, start again by turning the power off. Remove the switch from the wall and loosen two wires from terminals on only one side of the switch; reverse these wires. Retest; your new switch should work.

INSTALLING SPECIAL PURPOSE SWITCHES

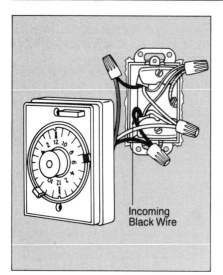

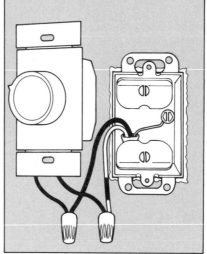

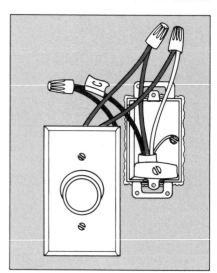

Installing a Time Clock Switch.
Time clock switches can only be installed in middle-of-the-run wiring where white neutral wires are available. With the power off, remove the old middle-of-the-run switch. Turn the power back on and, following testing and safety precautions on page 15, find the hot wire and mark it. Turn the power off and remove the wire nut from the white wires. Straighten out the bare end of the black wires.

Attach the incoming marked wire to the black wire of the switch with a wire nut. Also, using wire nuts, attach the other (outgoing) black wire to the red wire of the switch and the two white cable wires to the white wire from the switch. Mount the switch to the box, turn on the power and test the switch according to manufacturer's directions.

Installing a Single-Pole Dimmer Switch. Some single-pole dimmer switches are installed with screw terminals just the same as regular single-pole switches (page 30). Others, like the one shown here, have *leads*. Begin by shutting off the power to the circuit and testing to make sure that it is off. Next, using wire nuts, simply splice the leads to the wires within the box as you would a regular middle-of-the-run or end-of-the-run switch. Often, dimmer switches are large, so carefully arrange the wires in the box before pushing the switch into it; do not force it as it might break. If the switch does not fit properly into an existing box you will need to remove the box and install a larger one. If the switch you're using did not come with its own cover plate, simply reuse the old toggle-type switch plate. Attach it to the switch and place the rotary control knob on the shaft of the switch.

Installing a Three-Way Dimmer Switch. Before attempting this installation, it would be helpful to review the discussion of three-way switches (page 31). Three-way switches are used in pairs, but you can install a dimmer switch on only one, so decide which is the most used or convenient outlet to install it at. Begin by shutting off the power to the circuit and testing to make sure that it is off. As you remove the old switch, note its common terminal and mark the wire attached to it with masking tape. Attach this taped common wire to the common lead wire of the dimmer switch using a wire nut. Next, attach the remaining red leads of the switch to the traveler wires in the box using wire nuts. Carefully arrange the wires in the box; attach the cover plate and add the rotary control knob.

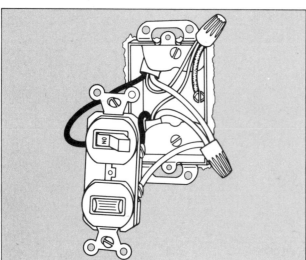

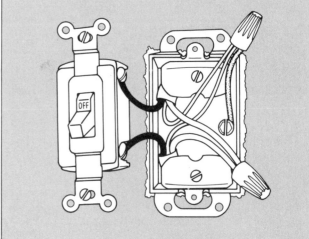

Installing a Pilot Light Switch— Separate Switch and Light Type.
Like time clock switches, these can only be installed in boxes where white neutral wires are available. With the power off, remove the old switch and then prepare the wiring as you would for a time clock switch (above). Most pilot light switches will have one silver-colored terminal and three brass-colored

terminals. Two of the three brass-colored terminals are joined by a brass strip. Connect the unmarked, outgoing black wire to one of these joined terminals and the other marked ingoing black wire to the brass terminal on the opposite side. Connect the white cable wires to the silver-colored terminal with a white pigtail. Attach with a wire nut.

Installing a Pilot Light Switch—Light In Toggle Type. These switches have only two brass-colored terminals. Simply connect the white cable wires together with a wire nut. Connect the black wires to the brass-colored terminals and then test your work. If the pilot light stays on when switched to OFF, reverse the black wire connections.

Fluorescent Light Dimmer Switches. Up until now, our discussion has been about incandescent light dimmer switches. But another variety, for fluorescent lights, is also available and can be quite useful for living areas.

Before considering this option, review the material on page 47 regarding fluorescent lighting. This will help you to determine what kind of fluorescent lighting you presently have.

To install a two-wire fluorescent dimmer switch, you must work with grounded fixtures that are fitted with 40-watt rapid-start lamps. If you have a 40-watt preheat or instant-start fixture, you will have to ground it first and then replace the lamp and lamp holders. Once you have done this, you can proceed with the steps shown here for installation.

The two-wire dimming system shown here is the simplest type to install; other kinds require additional wiring. One fluorescent dimmer switch can control up to eight 48-inch bulbs. However, you will need to replace the ballast in each fixture with a special ballast.

INSTALLING A FLUORESCENT DIMMER SWITCH

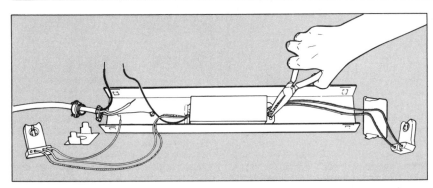

1 Begin by shutting off the power to the circuit and testing to make sure that it is off. Remove the lens, fluorescent tube(s), and metal cover. Remove the lamp holders at either end and pull them free. Disconnect the leads of the fixture from the house wires leaving the ground wire in place. Remove the locknut that secures the ballast, then carefully lift out the ballast, the lampholders, and the wires that connect the two.

2 Remove all the wires from the lamp holders except the short white wire that is attached to one of them. Often, these fixtures will have push-in type terminals instead of the screw type. If so, insert a thin nail or paper clip into the terminals to release the wires. Remove the old ballast.

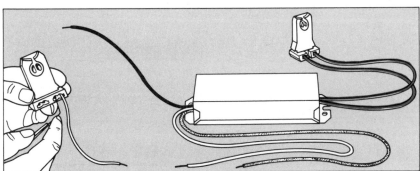

3 Attach the white wire from the new dimmer ballast to the lamp holder terminal with the short white wire in it. Connect the blue wire to the other terminal on the same lamp holder. To the other lamp holder, connect the two red wires from the ballast.

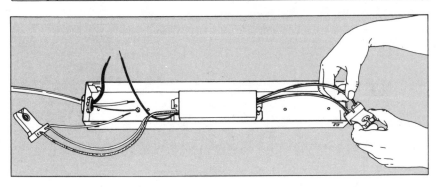

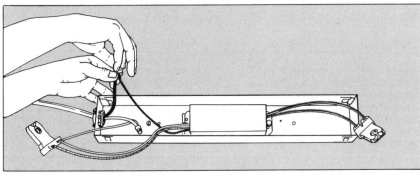

4 Secure the new ballast into the channel and make sure the locknut is tight so that the ballast will be grounded. With wire nuts, connect the black and white wires from the fixture to the incoming house wiring. Make sure the incoming ground wire is securely attached. When adding dimmers for several fixtures, new ballasts must be installed in every channel. Channels will be assembled as shown on page 69.

Choosing and Using Receptacles

Receptacles are installed and replaced according to what appliance the outlet will be powering and also by the available amperage of the existing or new wiring. Receptacle slots are designed to accept only plugs with corresponding shaped prongs based on their amperage rating.

Lower amperage receptacles are most commonly available as *duplex* receptacles, offering you two power sources at the same outlet. Higher amperage receptacles are usually available singly, as they usually supply individual appliances such as electric ranges or air conditioners.

Most homes have what is called *polarized* receptacles which means that one slot is larger than the other. Many appliance plugs have one larger prong so that when you attempt to plug into a polarized receptacle you must match the two neutral sides, thereby providing a grounding source.

In most modern homes you will also find that 120-volt, three-prong plugs are used exclusively. When making replacements in your home, be sure to use this type. Also, when purchasing them, check the important information on the receptacles to make sure that they are what you need.

Receptacles are easy to install and replace. There are two methods of wiring them—with *side-wired* devices, you attach the wires to screw terminals and with *back-wired* devices, you simply insert wires into push-in terminals. Some receptacles offer both methods.

Receptacle Combinations. Receptacles are available combined with other electrical fixtures. These dual-purpose installations are helpful in small areas, where it isn't practical to have many outlets. Shown on page 38 are a switch-receptacle combination and a light-receptacle combination with complete wiring instructions.

Receptacles for Safety and Convenience. Special receptacles make your home safer or neater. One safety model is the locking type which actually locks the prongs into the slots of the receptacle. For added convenience and a more pleasing look, use a clock receptacle. Either of these, shown on page 39, can be used to replace a standard 120-volt receptacle.

TWO KINDS OF WIRING ON RECEPTACLES

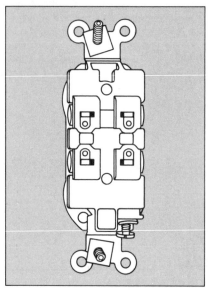

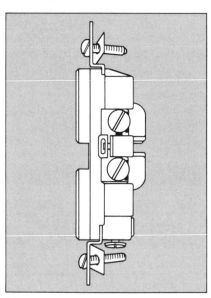

Back-Wired. Back-wired receptacles have push-in type terminals in their backs. They are wired using the methods on page 19. Often the word WHITE is printed on the side where you should insert the white neutral wires. The green screw at the bottom is for the ground wire. Some back-wired receptacles are also side-wired, offering you a choice of wiring methods.

Side-Wired. This receptacle has four terminal screws. Two brass-colored screws are on one side; they are for the black or red hot wires. On the other side are two silver-colored screws; they are for the white neutral wires. At the bottom is a green screw for the ground wire.

WARNING

When doing these wiring projects, first shut off the power and test to make sure it is off. Also follow the safety procedures on page 5.

MATCHING RECEPTACLES AND PLUGS

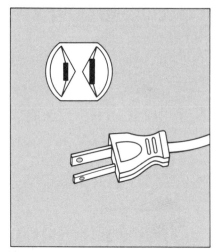

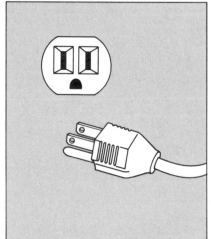

Ungrounded Two-Prong Plug (120 Volts, 15 Amps). Used on small appliances such as lamps, this receptacle has one slot larger than the other. If the receptacle is wired correctly, the larger slot is neutral; the smaller slot, hot. Plugs on some appliances are *polarized*, designed so that they can only fit into the slots one way thus matching the neutral and hot wires.

Grounded Three-Prong Plug (120 Volts, 15 Amps). The third prong indicates that the cord contains a grounding wire. When plugged into a correctly wired three-slot outlet, a ground fault occurring in an appliance or tool will be directed through the grounding wire to the grounding wire in the receptacle instead of through you.

REPLACING 120-VOLT RECEPTACLES

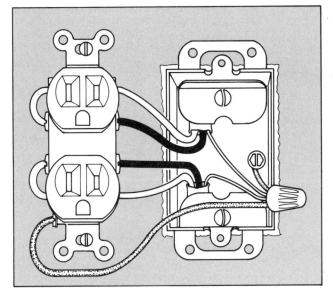

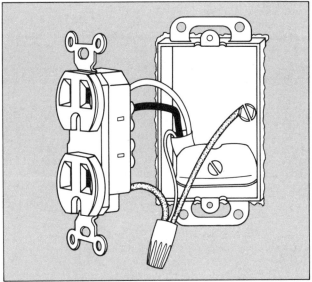

Middle-of-the-Run Receptacle. A middle-of-the-run outlet will be present if you find four wires—a black and white wire coming in and a black and white wire going out to the next outlet. There will also be grounding wires coming in and going out. To wire this receptacle, simply attach the two black wires to the brass-colored screws and the two white wires to the silver-colored screws. Then attach a short green wire to the green screw; attach another green wire to the screw in the box. (If there is no screw present, use a ¹⁰/₃₂ machine screw.) Join the two wires, along with the two bare grounding wires with a wire nut.

End-of-the-Run Receptacle. Only two wires will be coming into this box which is the last receptacle on the circuit. One will be black and the other white. There will also be a bare grounding wire. Connect the black wire to a brass-colored terminal screw and the white wire to a silver-colored screw. Attach a short green wire to the green screw on the receptacle. Next, attach another green wire to the box using the available screw or a ¹⁰/₃₂ machine screw. (**Caution:** Never use the screw that secures the cable clamp for this purpose.) Now join both of these green wires to the bare cable wire with a wire nut.

Note: These procedures are for nonmetallic cable. If your box is wired with armored (metal) cable, there will be no bare grounding wire present. In this case, attach a short green wire to the back of the box and attach the other end to the green terminal screw on the receptacle.

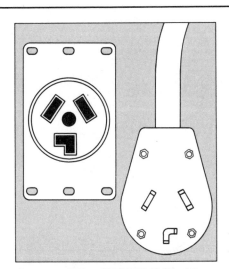

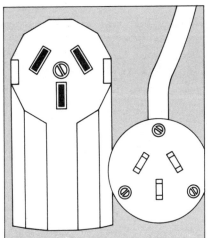

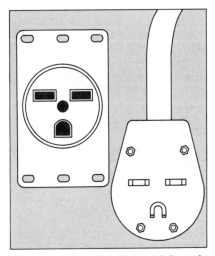

Grounded Plug (120/240 Volts, 30 Amps). Specifically designed for clothes dryers, this large receptacle provides 240 volts for the heating element and 120 volts for the timer and other mechanisms.

Grounded Plug (120/240 Volts, 50 Amps). Primarily used for electric ranges, the prongs are arranged at special angles to fit only a matching receptacle. The 240 volts provide power for the oven and burners at high temperatures; the 120 volts provide power for the burners at low temperatures and the range accessories.

Grounded Plug (240 Volts, 30 Amps). This plug is usually found on the cords of air conditioners. It will fit only into a matching receptacle.

INSTALLING A SWITCH-RECEPTACLE COMBINATION

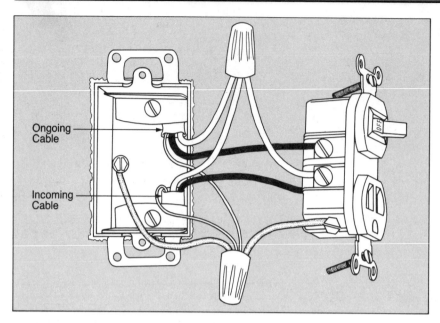

Ongoing
Cable

Incoming
Cable

This device can only be installed in middle-of-the-run wiring with incoming and ongoing cables. It replaces a regular switch and combines a single-pole switch with a receptacle that is always hot. Connect the incoming black wire to one of the brass-colored terminals that are linked by a metal tab. Connect the ongoing black wire to the brass-colored terminal on the opposite side. Attach a short white wire to the silver-colored terminal and connect it to the two white cable wires with a wire nut. Attach a short green wire to the back of the box with a machine screw and another green wire to the green terminal of the switch-receptacle. Connect both of these green wires and the bare grounding wires from the cables with a wire nut.

WARNING

When doing these wiring projects, first shut off the power and test to make sure it is off. Also follow the safety procedures on page 5.

INSTALLING A LIGHT-RECEPTACLE COMBINATION

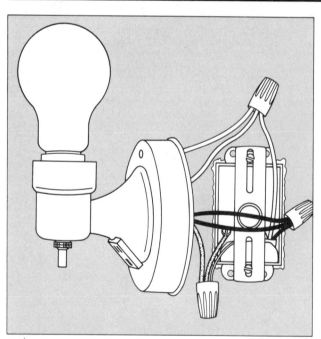

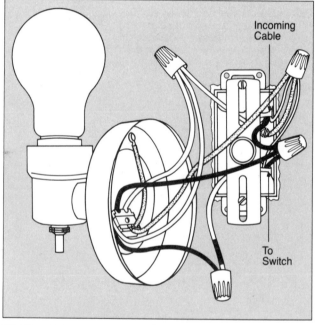

Incoming
Cable

To
Switch

End-of-the-Run Wiring. This installation replaces a regular light at the end of a wiring run, so there will be only one cable present in the outlet box. The switch that formerly controlled the light will now control both light and receptacle, but the receptacle will be usable only when the light is switched on. Connect the two black wires from the light-receptacle to the black cable wire using a wire nut. Attach all the white wires in the same manner. Attach a short green wire to the box with a $^{10}/_{32}$ machine screw; then connect this to the fixture's green grounding wire and the bare grounding wire from the cable—all three joined by a wire nut.

Middle-of-the-Run Wiring. There will be two cables present in the box for this installation. When completed, the switch will control only the light, and the receptacle will be usable at all times. Connect the black wire from the receptacle side of the fixture to the two black wires in the box. Identify the white wire going to the switch by marking it black, either with electrician's tape or paint. Connect this marked wire to the black wire coming from the light part of the fixture. Connect the two white wires of the fixture to the white wire from the incoming cable. Attach a small green wire to the box with a $^{10}/_{32}$ machine screw and then with a wire nut connect it to the green grounding wire from the fixture and the bare cable wires.

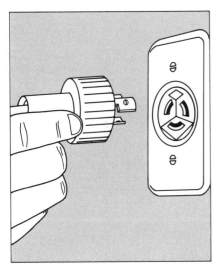

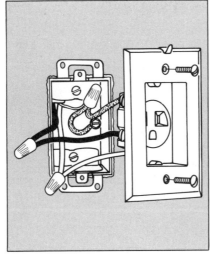

Special Receptacles for Special Uses. Replace any 120-volt receptacle with one of these specialized models that offers greater safety or convenience.

■ (L) **Locking Receptacle.** Useful for tools that are frequently moved about like drills and saws. The receptacle is styled so that it grips the prongs of the plug thus preventing it from being accidently pulled out. A corresponding locking plug must be substituted for the tool's regular plug.

■ (R) **Clock Receptacle.** Ideal for hanging an electric clock flush against a wall with no visible cord. The receptacle contains a hook at the top for hanging the clock and has a recessed area for inserting the cord.

WIRING HEAVY-DUTY RECEPTACLES

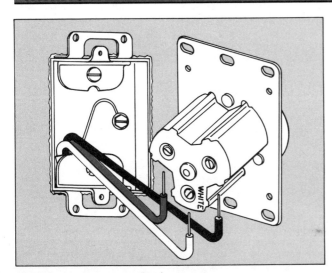

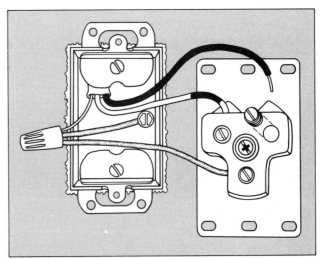

Wiring a 120/240-Volt 30-Amp Receptacle. Insert the white wire into the terminal marked WHITE and then tighten the terminal screw. Connect the black and the red wires similarly. Attach the bare grounding wire to the back of the box with a 10/32 machine screw.

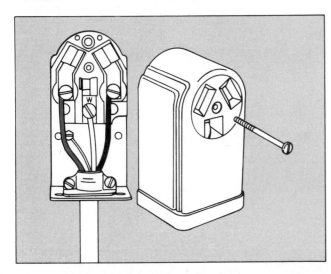

Wiring a 240-Volt 30-Amp Receptacle. The 240-volt cable should have only two hot wires—one black, the other white, plus a bare grounding wire. Be sure to mark the white wire with black electrician's tape or paint to identify it as a hot wire. Insert the black wire and black-marked white wire into the terminals with brass-colored screws. Tighten the screws. Cut two short pieces of green wire. Connect one piece to the green terminal screw of the receptacle and the other to the back of the box. Take the ends of these wires plus the bare ground wire and connect all three with a wire nut.

Wiring a Surface-Mounted Receptacle. 120/240-volt or 240-volt receptacles often come in surface-mounted models for use in utility areas where recessed wiring isn't necessary. Begin by removing the mounting screw and then follow the instructions for wiring according to the voltage and amperage rating of the receptacle, either 120/240-volt or 240-volt (left).

Easy Fixture Repairs and Replacements

Our lives are filled with electric appliances and gadgets. Some of them, when they are on the blink, must be taken to a repair shop. Others require a home service call from an electrician. But it's nice to know that not all problems have to be solved by someone else.

Many electrical problems, especially in lighting fixtures, can be remedied by a quick study of the problem, a trip to the hardware store, and a small amount of time invested in repair work. Replacing plugs can be a snap, and lamp repair, though it might seem complicated at first, is relatively undemanding.

Even more exciting are the simple replacements that you can make in your home. If you've never done it before, now is the time to replace an old ceiling fixture with a new one—you'll be amazed at how easy it is. Also, check the section on chandeliers and ceiling fans. Learning these skills will give you new inspiration for decorating and updating your home.

Throughout this chapter you'll notice that the wiring is standardized, color-coded, and easy to comprehend. In fact, most of the material presented here deals with basic installation procedures. Obviously, you'll save money by doing this work yourself. And, speaking of which, you might want to consider fluorescent lights in workspaces or other areas; they're more economical and aren't terribly difficult to install, replace, or maintain.

Simple Cord and Plug Replacements

Appliance cords and plugs are used repeatedly, and though we all try to follow good safety measures, they are sometimes abused and must be replaced. Fortunately, these are some of the easiest tasks for a homeowner.

Danger signs on cords and plugs include irregularly transmitted electricity, excessive heat, and physical damage. Usually a cord will wear out before a plug. If a cord is badly frayed or showing signs of wear, it is best to replace the entire cord with a new cord that includes a new-style polarized plug on its end. Polarized plugs, page 36, are now the norm as you will see when purchasing a new lamp or small appliance of any kind. Make sure that you purchase the same type and length of cord that is presently on the appliance. (A longer cord will cause more resistance and lessen the efficiency of the appliance.) Cords for appliances and lamps are *flexible cords*, named so because they are made of stranded and not the stiffer solid wires used in home circuits.

Most of these cords contain two wires; however, power tools and larger appliances will have a third grounding wire. Flexible cords like heater cords or power cords are encased in several light layers of insulation. *Zip, fixture,* or *type SPT* cord is used for lighting and other small appliances—it consists of stranded wires encased in a light thermoplastic insulation. Size 16- or 18-gauge is used for lamp wiring.

Replace plugs when their casings are cracked or if the prongs are so worn that they no longer connect well at the receptacle. There are basically two types —those that are self-connecting and those with screw terminals. The screw-terminal type plug, not shown here, is the old-fashioned round type that usually has a cardboard insulator over the wiring. This type is now restricted by the Code because the wires are dangerously exposed.

The self-connecting type plug is relatively easy to replace but it is recommended that you replace the entire cord and plug with a new polarized unit. These cord/plug combinations feature a

Attaching a Self-Connecting Plug.

Self-connecting plugs can only be used with zip cord. Do not separate the wires of the zip cord; instead, cut the end bluntly. There are two types of self-connecting plugs. For one, pull out the prongs and then push the cord in as far as it will go. Squeeze the prongs together tightly and push them into the cover. This causes a metal tooth inside the plug to pierce the cord. In the other type, the metal tooth works differently and the wire is inserted in the side of the cover instead of the back. Simply push down on the lever to make the connection. If the plug is polarized, make sure that the black wire connects with the tooth for the narrower of the prongs and that the white wire is pierced by the other tooth for the wider prong.

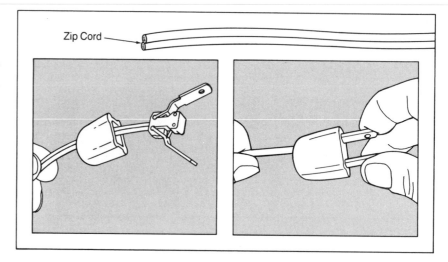

Zip Cord

REPAIRING A LAMP

This is the most most common type of lamp—a switch-and-socket combination. To diagnose the problem, check the bulb, cord, and plug. If they show no abnormalities, you may assume that the problem is in the socket or you can check further by testing (page 16). A faulty or outdated plug is easily replaced (above). Directions are given here for two procedures—replacing the socket or rewiring the entire lamp. Most often, the problem is in the socket. Begin by unplugging the lamp and removing the light bulb. Let it cool for awhile if it has been lit.

(page 16)

WARNING

Always unplug a lamp before working on it.

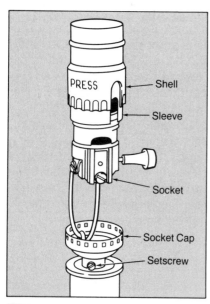

PRESS — Shell
— Sleeve
— Socket
— Socket Cap
— Setscrew

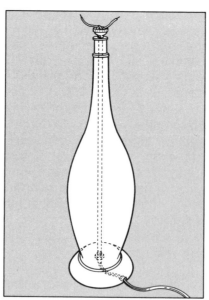

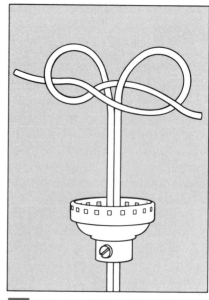

1 Start with the Socket. Remove the outer shell and the insulating sleeve by pushing at a point marked PRESS with a screwdriver. Loosen the terminal screws and pull out the wires to remove the socket assembly. Now, you have two options. If the socket cap is not damaged or corroded you should do the following: Attach a new socket, sleeve and shell by looping stripped and twisted wire around the terminal screws of the new socket. Connect the black wire to the brass-colored terminal and the white wire to the silver-colored terminal. Tighten the screws and fit the new shell and sleeve over the socket. Push the assembly down until it snaps into the socket cap.

If the socket cap is damaged and you wish to replace it, untie the knot in the cord, loosen the setscrew in the cap and unscrew it from the center pipe. Replace it.

2 Replacing the Cord. If you have not already done so, untie the knot in the cord resting in the socket cap. Pull the old cord from the bottom of the lamp; untie the knot at the channel. Now pull the entire cord free from the base of the lamp. Push the new cord through the channel in the base of the lamp and knot it, making sure that you have about four inches of cord coming out of the socket cap.

3 Tying the Underwriters' Knot. Carefully separate the cord, pulling them apart about two inches. Strip ½-inch of insulation off the ends of each and then tie them into an Underwriters' knot. Tighten the knot so that it nestles in the neck of the socket cap. Reassemble the socket, using the instructions in step 1.

molded plug which assures you that the plug wires cannot be exposed. This is a safety measure against electric shock caused by a 'hot' lamp shell.

When purchasing a plug, look for the polarized type, after making sure that your receptacles can accept them. The Code requires this type for all replacements because, wired correctly, they polarize the appliance through the neutral side of the circuit. If your receptacles cannot accept them, consider installing new receptacles, an easy replacement project.

Fixing a Lamp—Getting the Bulb to Glow Again

Lamps, portable incandescent light fixtures, have a practical and decorative value different than ceiling fixtures. Lamps tend to last a long time but if you use them often, they will succumb to wear. Perhaps you have an attractive old lamp that is in need of repair, or maybe a lamp that you're using flickers occasionally. The following repair skill will take care of those problems plus it will give you the basis for constructing your own lamp out of a favorite material if you someday choose to do so.

Though lamps come in different shapes and sizes, they all have the same basic working parts. Whether they're used on tables or floors, lamps have these four components: socket(s), switch, cord, and plug. The most common type has the socket and switch combined in one mechanism.

In this type, the switch-and-socket combination, there can be various kinds of switches built into the socket, such as the popular rotating switch or the more outdated push-in or pull-chain types. Lamps sometimes have two or more sockets and these may or may not have their own switches.

Another type of lamp has the switch and socket separated; often the switch will be located at the base of the lamp. If you are having trouble with such a lamp, you should replace the switch; usually the problem is there.

In troubleshooting lamp problems, if you can't visually identify problems in the plug or cord, you can assume that the problem is in the socket/switch or you can check with a continuity tester for more detailed information (page 16).

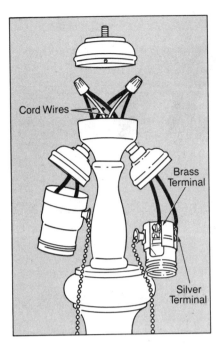

Repairing a Lamp with Two Sockets. This type of lamp has two separate sockets, each with its own wiring and switch. The socket wires are connected by wire nuts to the main cord wire coming from the center tube of the lamp. To replace the sockets, follow the same procedure that was used for repairing a single-socket lamp, (left, step 1). To replace the cord, take off the top cover and then remove the wire nuts. Replace the cord wire as for a single-socket lamp (left, step 2), and pause after you have stripped ½-inch insulation from the cord ends. Now you must connect the wires coming from the brass-colored socket terminals to one of the main cord wires and the wires coming from the silver-colored socket terminals to the other main cord wire.

Lamp repair is not difficult; it is simply a matter of replacing one of the faulty components. When you know that you're using a good bulb in a lamp and it will not light, unplug it and pry up the small brass contact in the socket base. Try again. If it still will not light, or if the cardboard insulation has deteriorated, replace the socket.

Sockets are easy to hook up; even a three-way bulb socket has the same wiring as that of a standard one-way socket, although it will require a heavier gauge wire. When purchasing sockets, be sure to specify one-way or three-way, as they look very similar.

If a lamp shows problems at either the cord or plug, go ahead and replace both with a new molded type plug and, preferably, a polarized unit. If you were to shop for a new lamp you'd see that this is a safety requirement for all new fixtures.

Updating and Highlighting Rooms by Replacing Lighting Fixtures

One of the simplest ways to make a decorating changeover in a room is to replace the lighting fixture. Modern fixtures come in such a great variety of styles; there is truly something for everyone and every type of room. Another reason for changing fixtures, of course, is that they are not working properly. In order to test the wiring at a lighting outlet, use a voltage tester (page 14).

Fixtures are available for both ceilings and walls, and they come in various shapes and sizes. The very simplest type of lighting fixture is mounted by simply screwing it into the box via the small mounting tabs on the box. Perhaps you have seen this type of fixture, usually porcelain, in garages or basements.

The fixtures shown on the next page, however, require additional hardware to give them extra support. You can begin by removing your old fixture and examining the mounting hardware already available in the box. If your new fixture is of the same size, shape, and weight as the old one, you can probably reuse it. If not, before buying parts, you can try to determine specifically what hardware you will need by studying these diagrams. Or, you can purchase a variety package of hardware for lighting fixtures.

The process for changing fixtures is very logical: first remove the old fixture; next, wire the new fixture; and last, mount it. Removing old fixtures can be tricky. Check for screws around the edge or a disguised cap nut in the center. Another method of removal is to turn the cover plate to the left to unscrew it. The wiring is essentially the same for all fixtures, but the mounting methods vary.

REPLACING LIGHTING FIXTURES

Shown here are various mounting procedures. In each instance, begin by turning the power off at the service panel. Next, remove the old fixture and attach it to and hang it by a wire coat hanger so that you can easily disconnect the wiring. *Do not let the fixture hang by its wires.* After detaching the wires from the fixture, strip away ½ inch of insulation from the wire ends. Making the connections is very easy—black wires connect to black, and white to white. Do not disturb any other wires in the box.

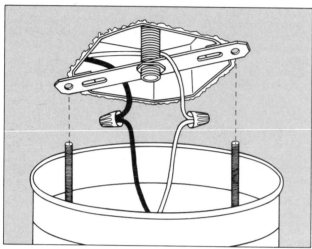

Mounting with a Strap and Stud.
When a fixture is too large to attach to the screw holes in the tabs of the box and the box has a stud projecting from its middle, you must use a mounting strap. Screw the strap onto the stud and then secure it with a locknut. After making the wire connections, carefully fold them into the box, and then screw the fixture to the holes in the mounting strap.

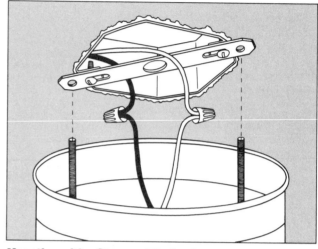

Mounting with a Strap and No Stud.
When the box contains no stud, you must attach the mounting strap by inserting screws through the strap slots and into the box tabs. Make the wire connections, fold them into the box, and then screw the fixture to the holes in the mounting strap.

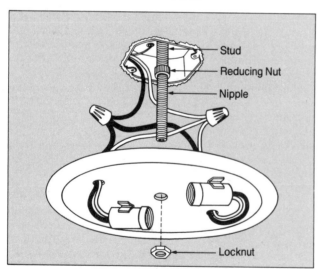

Stud
Reducing Nut
Nipple
Locknut

Mounting a Center-Mounted Fixture with a Stud. In order to connect the central stud in the ceiling box to the threaded nipple which will support the fixture, you must use a reducing nut. It has two different threaded ends to accept both the stud and the nipple. Screw it into the stud and then screw the fixture nipple tightly to it. Connect the black wires of the fixture to the black house wire using a wire nut. Connect white fixture wires to the white house wire with a wire nut. Mount the fixture base and secure it with a locknut.

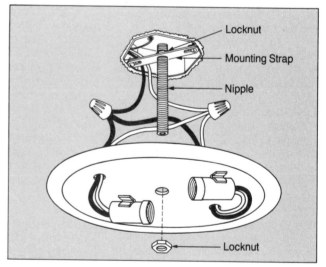

Locknut
Mounting Strap
Nipple
Locknut

Mounting a Center-Mounted Fixture with No Stud. If a box has no stud, you need to use a mounting strap for its center hole. Fasten the fixture nipple through the center hole of the mounting strap with a locknut. Attach the strap to the box tabs with screws. Mount the fixture base and secure it with a locknut.

Mounting a Heavy Chandelier or a Ceiling Fan and Light Combination. If a chandelier weighs less than ten pounds, it can be mounted with a nipple and a strap attached to the tabs of the ceiling box (previous page, bottom right). But if it weighs more, you will need a hickey and you will have to work with a ceiling box that has a stud. If your box has no stud, you must replace it with a box that has one.

If you are installing a paddle-type ceiling fan, with or without lights, you will need to take extra measures to provide adequate support. Special ceiling boxes are made specifically for these fixtures; for any fixture weighing over 35 pounds additional independent structural support must be provided. In short, a box is not enough. As always, follow the manufacturer's instructions when making such an installation.

In a ceiling fan/light installation the wiring is like that for the chandelier.

However, you may wish to add an additional wire coming from the wall switch. That way, if you want to add a dimmer switch you will be able to control the brightness of the light without affecting the speed of the fan.

Economical and Efficient Fluorescent Lighting Fixtures

The main benefit of fluorescent lighting is the great cost savings for the homeowner. A 40-watt fluorescent tube puts out six times more light than a 40-watt incandescent light bulb and lasts approximately five times longer.

On the other hand, fluorescent lights have a bad reputation for providing unnaturally colored or harsh lighting and consequently most people don't prefer them. Today, however, technology has given us a whole range of 'white' tubes to choose from … cool, warm, and soft. Each of these intensifies warm or cool colors and blends to varying degrees with incandescent colors. The new 'deluxe' fluorescent tubes most closely approximate incandescent lighting, but they produce 30 per cent less light, per watt, than other types.

Because fluorescent tubes are limited in design to long tubes or circles, they are often used in utility areas, over workbenches, or recessed into ceilings. They also can be used to provide decorative touches when installed in cornices, coves, soffits, or valances in living areas (page 68).

Fluorescent fixtures have two basic designs. The starter will either be separate from the ballast or built into it. When troubleshooting the first type, always replace the starter first, as it is an easy process. The second type might require installation of a new ballast; the cost of this expensive part often comes close to the cost of an entire new fixture.

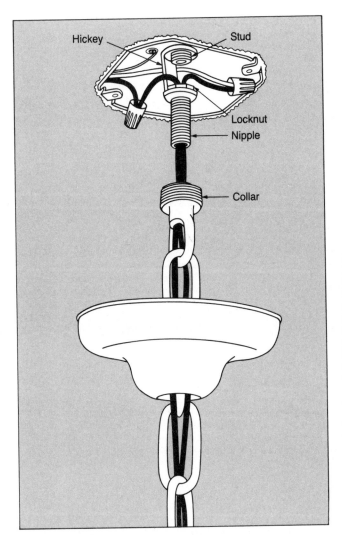

WARNING

■ When mounting a chandelier or heavy fixture, use a helper. Never let a fixture hang by its wires.
■ When doing these wiring projects, first shut off the power and test to make sure it is off. Also follow the safety procedures on page 5.

Installing a Heavy Chandelier. Begin by removing the old fixture and preparing the wiring according to the instructions for replacing lighting fixtures (left, top). Screw a large adapter called a 'hickey' onto the stud and attach a nipple to it. The nipple should be screwed into the hickey far enough so that when the chandelier's collar is in place it will hold the canopy flush to the ceiling. Determine this distance and then secure the nipple with a locknut.

Pull the wires up through the collar, canopy, nipple, and the lower part of the hickey. Connect the wires of the chandelier to the house wires with wire nuts—hot wire to the black wire and neutral wire to the white wire. Secure the collar to the nipple and the canopy to the collar, as shown.

Note: Fixtures vary as to how these components are connected. Sometimes, the canopy will be located between the nipple and collar; in this case, screwing the collar to the nipple will anchor the canopy in place.

Troubleshooting Fluorescent Lighting Fixtures. First, determine which kind of fixture you have—rapid-start, instant-start, or starter type. Check this list for the current problem with your fixture and then follow the steps, start to finish, testing after each step. Always turn the power off at the service panel when doing repair work.

■ **Tube won't light.** Check the circuit. If tube is black at both ends, replace. If not, replace starter. If it still won't work, replace ballast.

■ **Light blinks off and on.** Straighten and sand the tube pins. Turn off power; clean and lightly sand contacts in the light unit. Replace the starter. (If light is low temperature, below 50°F, use a cold-weather starter.) Replace the ballast.

■ **Light flickers.** A normal function of a new tube. For an old tube, first replace the tube. Replace the starter. Replace the ballast.

■ **Discoloration of tube.** Brown or gray bands at tube ends are normal. Expanding black bands at tube ends indicate burnout; replace tube. Black bands on new tube—replace starter. Replace ballast.

WARNING

When doing these wiring projects, first shut off the power and test to make sure it is off. Also follow the safety procedures on page 5.

INSTALLING A ONE-TUBE FLUORESCENT FIXTURE

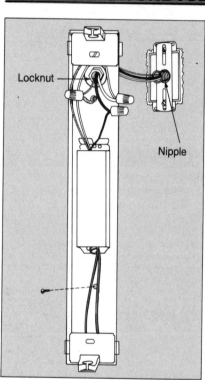

Locknut

Nipple

To a Wall with a Strap. Begin by shutting down the power at the service panel. Remove the old fixture from the box and then attach a nipple to a strap. Screw the strap to both of the box tabs. Remove the lid of the fixture and then after determining how you want it positioned on the wall, punch out one of the perforated knockouts in the channel with a screwdriver to create a hole. Pull the house wires through the nipple and the knockout hole and slip on the washer and locknut. Connect the black and white house wires to the black and white fixture wires, matching colors. Also connect the green or bare grounding wires. Note: Depending on the length of the fixture wires, either make the connections in the channel (as shown here) or in the box. Tighten the locknut securely. If the center channel hole was used, the locknut will support the fixture sufficiently. If the box is mounted at either end of the channel, the other end must be supported with an additional screw.

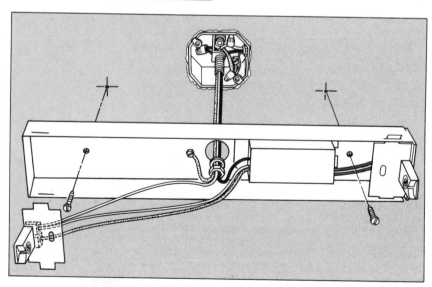

To a Ceiling with A Stud. If the box you are using has a stud, attach a hickey and a nipple to it (page 44). Then follow all of the basic instructions above for mounting a fixture to a wall with a strap. For additional support, screw the fixture channel to the ceiling, as needed.

FLUORESCENT LIGHTING FIXTURES

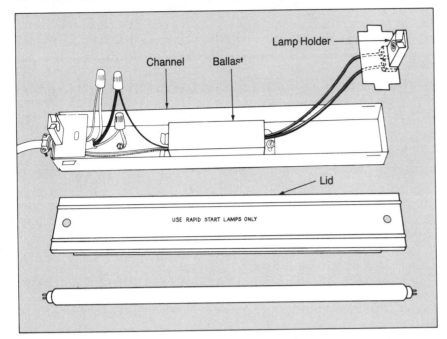

As shown here, there are three basic kinds of fluorescent lights. In the most popular kind, the rapid-start, the starter is contained in the ballast and there is no delay when you switch on the light. In the instant-start type, a high voltage is needed to charge the tube which is distinguished by its single pins at each end. In the older-style starter type, a separate starter powers the tube and there are several seconds of delay when the light is switched on.

Rapid-Start. A common source of problems for this type of fixture is improper grounding. The tube should be no more than ½ inch from a grounded metal strip—either the channel or a metal reflector. Another source of problems can be usage. If this light is turned on often, but only for brief periods of time, it will interfere with key mechanisms and the result will be a slow-starting light. If you do succeed in lighting such a fixture, leave it on for several hours and the problem might cure itself. If this doesn't work after several tries, the problem is elsewhere.

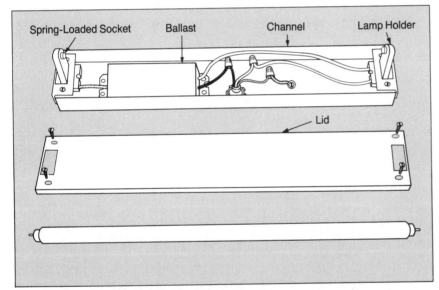

Instant-Start. This model is relatively trouble-free so when it does not operate first check to see if the tube is connected properly. One side of the holder has a spring-loaded socket which acts to shut off the current if the tube pin is not fully inserted; so a quick jiggle might remedy the problem. If a bulb is partially blackened and the light produces brilliant orange flashes, you should replace it immediately. This indicates that the high voltage of the fixture is still operating and allowing it to continue could damage the ballast.

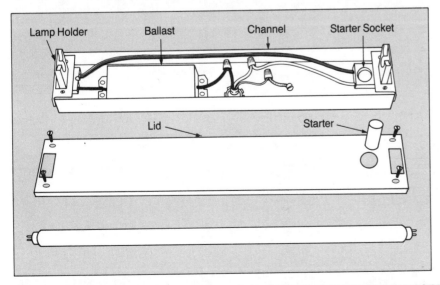

Starter Type. This fixture contains a small starter located in a socket near one of the lamp holders. Replace the starter with one of the same wattage that is printed on the tube. Begin by turning off the power and removing the tube. Pull the starter from its socket by turning it counterclockwise. If there are wire nuts connecting the starter to the socket, remove them. Reverse the process to install a new starter.

Planning a New Circuit

This is a 'preparation' chapter, designed to give you all the information that you'll need before moving on to the advanced projects of installing new branch circuits, extensions, and boxes. Up until now, you've been in the 'elementary class' of wiring; next you will graduate, but until then you need a little 'secondary schooling'.

Here you will see the big picture of how power is transmitted into your home and distributed at your service entrance panel through branch circuits. You'll learn about the different kinds of branch circuits, how they are distinguished by their loads, and how to calculate a load. This will help you to figure out whether to add an entirely new branch circuit or simply one or two outlets at the end of an existing circuit. A

sample map is shown to give you an idea of how most homes are wired and to illustrate how you can map out your own circuits.

Your service entrance panel plays a vital role in this power transmission. Whether yours is a circuit breaker type or a fuse box, you will need to record your circuits at the box for future reference. Circuit breakers are discussed briefly; however, major skills relating to them appear in other sections of the book.

The chapter ends with a step-by-step list of the installation process. Although the list itself looks long, once you read through it you'll see just how clean-cut and easy this process is going to be. There is also a structural drawing of a house. A brief study of it will prepare you for what's to come in the chapters on running cable.

An Overview of Home Circuitry

In order to plan for a circuit, you must first understand the present circuitry in your home. In other chapters, we've discussed the elements of circuits—the switches, receptacles, and so forth—but it helps to see how they all function together before moving along to the installation process.

Electrical power originates at a power plant. From there it travels through various transformers until it arrives close to your house via overhead wires. (Some people receive their power through underground wiring but most people have this type.) Outside of your

house there is a point where the utility lines connect to your house wires. This is called the *service head*. From there, wires travel down to the electric meter which has been installed by your utility company to measure your power usage. Finally, power enters your home and is connected at the main service equipment panel board (here called the service entrance panel).

Since the 1950's, most new homes have been supplied by three wires coming from the service head and into the service panel. Two of these wires carry 120 volts each and the third wire is neutral, at zero voltage. The neutral wire is connected at the service entrance panel

to a ground wire and the grounding process is further carried on (page 3).

The two 120-volt wires supply a home with 240-volt service. Individually, 120-volt wires are sufficient to supply circuits for lighting and general purpose use. For heavy-duty use, such as that needed to power a water heater, both of the 120-volt wires are combined to give 240 volts. Wires leaving the service entrance panel and branching throughout the house are referred to as *branch circuits*, although for brevity's sake, they may be referred to in this book simply as 'circuits'.

Older homes might still be supplied with only two incoming wires—a 120-

TYPICAL HOME CIRCUITRY

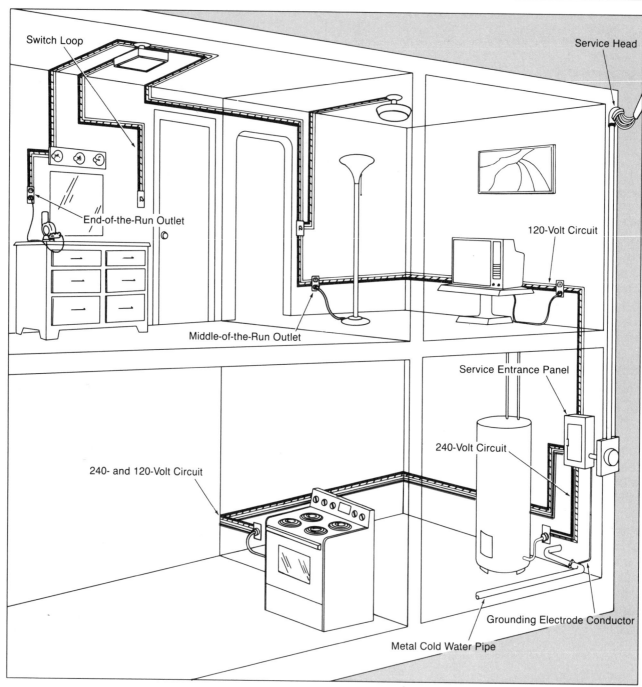

Switch Loop

Service Head

End-of-the-Run Outlet

120-Volt Circuit

Middle-of-the-Run Outlet

Service Entrance Panel

240-Volt Circuit

240- and 120-Volt Circuit

Grounding Electrode Conductor

Metal Cold Water Pipe

Though it only shows three circuits, this draw-ing illustrates both the varying voltages of home circuitry and also the various wiring patterns that you should know if you are going to install new circuits. This house has the standard 120/240 volts of most modern homes. Coming from the service entrance panel is a grounding electrode conductor which is connected to a cold water pipe. Every circuit in the house contains a bare grounding wire that connects at the service entrance to this system.

Downstairs, there is a 240-volt circuit for the water heater; it is identified by the two 120-volt hot wires, one black and one red (the second hot wire may instead be white, remarked as a black wire). The two hot wires alternate for the return in the circuit. A similar

set of wires goes to the outlet for the electric range, but this also contains a white wire. The black and red wires transmit the 240 volts necessary for the range's heating ele-ment. The white wire powers its timer and light, and a white wire provides the return route to the service panel.

Upstairs, where lower voltage is needed, there is a 120-volt circuit supplying outlets for lights and appliances. Typically, these circuits have up to ten outlets and span several rooms. Every outlet on the circuit requires middle-of-the-run wiring except the last outlet—in this case, a receptacle giving power to the hair-dryer in the bedroom.

Sometimes, as shown at the television outlet, middle-of-the-run wiring has one set of wires entering the outlet and the same set

exiting and going on to the next outlet. At other places, the wiring will branch off, such as it does in the first switch outlet on the cir-cuit (wires lead to the living room overhead light and another set goes on to the next outlet). Three or more cables may be con-nected in this manner. The next outlet is another overhead light, in the bedroom. This one has three sets of wires—the third set ending at the switch. In this configuration, called a switch loop, power is carried only to and from the switch. The white wire in this box is remarked as a black wire since it is hot. Finally, there is the last outlet on the cir-cuit, for the hairdryer. In this end-of-the-run outlet there is only one cable for the supply and return of the current.

volt and a neutral—and thus will lack the power to supply certain appliances. Though modern homes have circuit breakers at the service entrance, some old homes still have fuse panels which lack certain advantages and are not as easy to operate.

Identifying Circuits at the Service Entrance Panel. Your service entrance panel should have a list posted on its door that tells by number which breaker or fuse protects which part of your house. If it doesn't clearly state this information, then you and a helper can attain it and make your own list.

First, on a sheet of lined paper, number all of the positions for breakers or fuses. Then, throughout your house, turn on all the lights and plug in small lamps or appliances to receptacles (do not concern yourself with the major appliances at first). Next, switch a breaker to off and have a helper check the house, upstairs and downstairs, for which outlets are not working; record this on your list. Two story houses are almost always wired with circuits expanding two floors so that one whole floor will not be powerless in the event of a short circuit or other trouble. Continue switching the breakers off and recording the information on your list.

Ampacity of Circuits. The total amount of current required in all parts of a system is called the electrical *load*. Different kinds of wiring and different size wire is required to carry the load in each area of the home. Each circuit is protected by an overcurrent device—either a circuit breaker or a fuse. If a circuit becomes overloaded, or if a short circuit develops, the device will stop the flow of current.

Homes have many branch circuits. There are two reasons for this. If only one circuit were used for the whole house, the wires would have to be of a thicker gauge, and thus more expensive. Another reason is safety. Everything electrical in your house would fail in the event of a problem in the system if it were all on one circuit, so it makes sense to have more than one.

Calculating a Load on a Circuit

As an aid, circuits are divided accordingly: *General purpose* is defined as the smallest load in the house. It carries current for lighting fixtures, radios, stereos, televisions, and other low-voltage type appliances. *Small appliance* circuits carry loads larger than general purpose circuits but much smaller than individual appliance circuits. Typical small appliance circuits are those powering

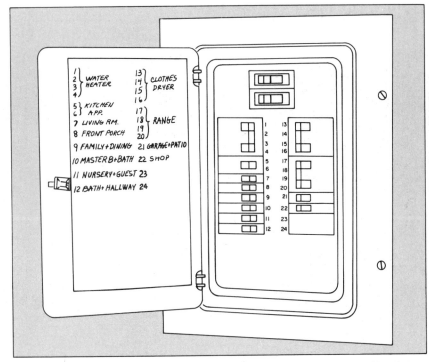

Identifying Circuits at the Service Entrance Panel. This service entrance panel is the circuit-breaker type (see page 23 for a fuse-type panel or box). Most panels will have a list on the cabinet door identifying the 'branch' circuits. The main switch at the top is a 100-amp breaker which prevents all of the branch circuits from drawing excess power from the power company's line. If you switch it to OFF you will shut down the circuitry in your home but you will not affect the main circuit or power line coming into the box.

Note on the list that the number of breakers needed for each appliance is indicated as well as the name of the appliance. Linked double-pole breakers are used for 240-volt circuits to the water heater, clothes dryer, and the range. Single-pole breakers protect 120-volt circuits in various areas of the house. There are empty spaces in the panel for the addition of new breakers and at the bottom of the list for writing in the location of your new circuit(s).

shop equipment, space heaters, irons, refrigerators, food processors, coffee makers, and the like—generally for kitchens and utility rooms. *Individual appliance* circuits are needed for major appliances such as clothes dryers, water heaters, ranges, dishwashers, garbage disposals, and air conditioners.

Calculating a load can be very tricky because of two factors. First, it seldom occurs that all appliances will be in operation at one time on a circuit. A good example of this is an air conditioner on the same circuit as a heater. Second, it seldom occurs that appliances are utilizing their full power all of the time. For example, a self-cleaning range is most generally used for everyday cooking.

The *National Electrical Code* specifies that you must use the maximum capacity size cable and circuit breaker when wiring for appliances but the Code also permits you to allow for these diversified demand factors when you calculate the total required of the service entrance. Though information is available for making this calculation,

the easiest method is to obtain the advice of a qualified electrical inspector.

A Guide for Planning Your New Branch Circuit or Circuit Extension. First, determine what kind of circuit or circuit extension you are going to install by choosing from the three types mentioned above: general purpose, small appliance or individual appliance. Then follow the instructions below to calculate the load.

■ **General Purpose Circuits.** The Code requires three watts of power per square foot of floor space but it does not limit the number of fixtures or receptacles that these circuits serve. Generally, a 20-amp circuit should not supply more than sixteen outlets or a 15-amp circuit more than twelve. Stationary appliances such as refrigerators may be supplied, but the sum of appliance ratings must not exceed half of the circuit rating and no single appliance can exceed 80 per cent of the circuit rating.

■ **Small Appliance Circuits.** Single-family dwellings should have at least two of these 20-amp circuits; most

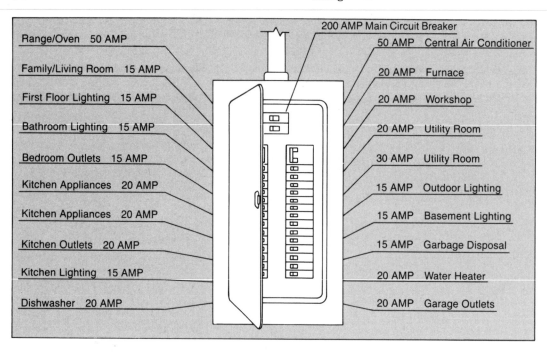

Range/Oven   50 AMP
Family/Living Room   15 AMP
First Floor Lighting   15 AMP
Bathroom Lighting   15 AMP
Bedroom Outlets   15 AMP
Kitchen Appliances   20 AMP
Kitchen Appliances   20 AMP
Kitchen Outlets   20 AMP
Kitchen Lighting   15 AMP
Dishwasher   20 AMP

200 AMP Main Circuit Breaker
50 AMP   Central Air Conditioner
20 AMP   Furnace
20 AMP   Workshop
20 AMP   Utility Room
30 AMP   Utility Room
15 AMP   Outdoor Lighting
15 AMP   Basement Lighting
15 AMP   Garbage Disposal
20 AMP   Water Heater
20 AMP   Garage Outlets

Ampacity of Branch Circuits. Shown here are typical branch circuits for a family dwelling and the amperage for each. When adding new circuits, remember that the circuit breaker should be of the same amperage as that of the cable you are using.

homes have more. Circuits generally supply two or three receptacles and these must be used for high-wattage appliances such as those used in the kitchen. These circuits are not permitted to serve receptacles or lighting fixtures in any other room. To calculate a load, use the formula for general purpose circuits given above.

■ **Individual Appliance Circuits.** Only one individual appliance circuit is required by the Code—a 20-amp, 120-volt laundry circuit for a washing machine. Most homes, however, have many others—for clothes dryers, freezers, ranges, air conditioners, and other high-wattage appliances. These circuits can be powered by 120 or 240 volts and they can be wired directly to an appliance or to a receptacle within six feet of the appliance. Each circuit

supplies only one or sometimes two appliances. The Code does not specify the amperage for such circuits, but the capacity of their circuits must be greater than the amperages indicated on the appliance nameplates.

For motor-driven appliances such as a stationary power tool, the circuit amperage is generally figured at 125 percent of the full load current rating of the motor. To be sure, however, always refer to the equipment instructions and follow them specifically. This is very important to allow for the current surge when the motor starts.

Determining the Cable and Estimating the Wire Size

Once you have calculated the load you can figure out the size of wire, type, and

number of cables to use. Use the information on page 9 plus the chart shown here.

Wire Size and Overcurrent Protection

Type of Circuit	Copper Wire Size (AWG)	Overcurrent Protection
General Purpose	14	15
	12	20
Small Appliance	12	20
Individual Appliance	10	30
	8	40
	6	55

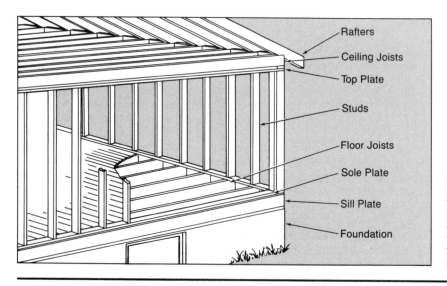

Rafters
Ceiling Joists
Top Plate
Studs
Floor Joists
Sole Plate
Sill Plate
Foundation

Anatomy of a House. If you haven't done much home repair work in the past, you might not know the language of house structure that carpenters use when they build a house. In this skeletal representation, you can see all of the wooden structural parts where wiring can be installed. Familiarize yourself with the terminology so that you will understand clearly the location of wiring runs.

The Steps to Take When Installing New Circuits

No matter how extensive your wiring project is—putting in only one new box or installing new circuits—here are the basic steps that you will follow. Read through this now while planning and then refer to it later as you actually do the work.

1 Determine where your new box(es) will be located (pages 75 through 83). Try to find nearby studs and joists and then break through the wall or ceiling. If it's in a ceiling, figure out which way the joists run.

2 Find an existing box near the new one to provide power for the new wiring run. Try to find a direct route. This may be difficult, especially in ceilings; if not possible, use an alternate route (page 57).

3 Shut off the circuit's power at the service panel. Remove the cover plate of the existing box and then check it inside to see if you can connect new wires to existing wires or terminals (page 55).

4 Count the connections in the box to determine if there is enough room for more wires (page 74). If the box is too small, you will have to install a deeper box, gang it with another box, or start your circuit from another place.

5 Check the circuit breaker or fuse at the service entrance panel to determine whether the existing circuit can accept the additional load that your new branch circuit will put on it.

6 Make a map of the room and adjoining rooms showing the existing box, new box, location of studs and joists, and the desired route for running the new cable.

7 Make a list of the materials you will need: new box or boxes, hardware, cable, and clamps. Use the guide on page 119 to make your shopping list.

8 Run the new cable from the existing box to the hole for the new one. Usually the most demanding part of the job, this may involve cutting access holes in walls and ceilings, drilling holes in beams for pathways, and snaking the cable (pages 57-63.)

9 Shut off the power to the existing box and clamp the cable to both the new box and the existing one. Usually the old box will require an internal clamp, whereas the new one will accept either an internal or an external clamp.

10 Mount the new box according to the composition of the wall and its proximity to a stud or joist (pages 75-83).

11 Connect the wires correctly from the cable in the existing box to wires and terminals. Then connect the wires in the new box.

12 Test the new circuit branch with a continuity tester. At the last box on the run, place one probe of the tester on the black wire. Place the other probe in these three positions: to the white wire, the grounding wire, and the box. The bulb should not glow in any position. If it does, you have a short circuit and you must check each box for poor connections or bare wires touching each other. Also check the cable for frayed insulation. Make any necessary corrections.

13 Test with a voltage tester, this time with the power turned on at the service panel.

14 Shut down the power and patch up the holes that you have made in the walls or ceilings (page 82). Attach faceplates and turn on the power—your new branch circuit is ready for use.

A MAP OF HOME CIRCUITS

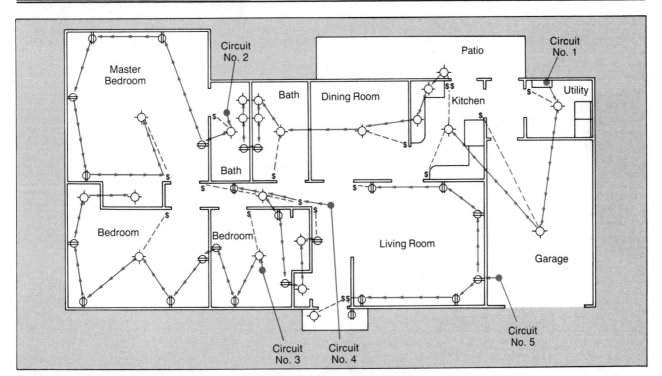

Here is a simplified drawing of five house circuits. Shown are only the "general purpose" circuits controlling lights and receptacles for small appliances. Identified by symbols are lighting outlets, switches, and receptacles. The broken lines indicate which light is controlled by a switch. The red dots show the beginning outlet for each circuit. Though the cables aren't shown, each of these circuits originate at the service entrance panel. Note that wires often travel from room to room; in a two-story home they travel from floor to floor.

$ Switch	**⌽** Receptacle
Ceiling Light	Wall Light

Installing a New Circuit

Although the previous chapter gave you planning information for installing a new circuit, don't just choose a project in this chapter and leap in. It's recommended that you read all of the material first to see what your options are. You can make your new circuits or circuit extensions extremely elaborate or as simple as possible.

Here you'll learn how to tap into various kinds of boxes to extend circuits. Then you'll see the ups and downs of running cable—coming from basements and attics, snaking through walls, going around doorways and behind baseboards. If you were to install wiring into the framework of a house before the walls were covered, you would recognize the simplicity of the job. But now, you're probably a bit timid about 'tearing apart' the house. It's only natural to feel awed by

the mystery of the process—and yet, a well-paid electrician would only do the same. So if you're willing to do the patching (page 82) or to pay someone else to do that end of the job, you have nothing to lose by doing the wiring yourself.

If you want to install new circuits for lighting, study the sample cable runs in this chapter, plus the many step-by-step fluorescent fixture installations. These projects, done carefully and correctly, can add greatly to the beauty and value of your home.

Lastly, the chapter ends with some switch and receptacle options that can give you even more ideas for updating or problem-solving. Now that you understand the principles of circuitry and have gained some hands-on experience making installations, you're bound to start thinking creatively about your own unique home wiring.

Preparing for the New Circuit or Circuit Extension

Deciding whether to add an entirely new circuit or to simply tap into an existing one can be difficult. As a general rule, however, don't do any more work than is necessary. Make sure that you fully understand the steps listed on page 53. Then consider what your load requirements are going to be (page 51). If you only need a few outlets for low-wattage appliances or lighting outlets, you can usually get by with extending a circuit from an existing outlet; and often it is easier to install several extensions rather than adding a new circuit.

If you do choose to add a fresh new circuit, you must install cable at the service entrance (page 22) and you should also contact your local building department for restrictions or requirements.

The Requirements for Extending a Circuit

If you decide to make a circuit extension from an existing outlet, there are several requirements that must be met. First, check the outlet wiring at the box that you intend to tap.

Understand that you cannot extend a circuit from any outlet. In the box that you use there must be both a hot and a

neutral wire that are in direct connection with the power source at the service entrance. In other words, you cannot use a box with end-of-the-run wiring. You can, however, tap into any of the five types of outlets shown on page 56.

Now, measure the box that you're going to tap into. Count the number of conductors that will be included in the box and then check the chart on page 74 to make sure that you aren't going to overcrowd it with wiring. Also check the box for a knockout. Remember that you can always change the box if you need to and this is often preferable to running cable on a more difficult route. (Boxes are covered on page 75.)

CONNECTING NEW CABLES IN OLD BOXES

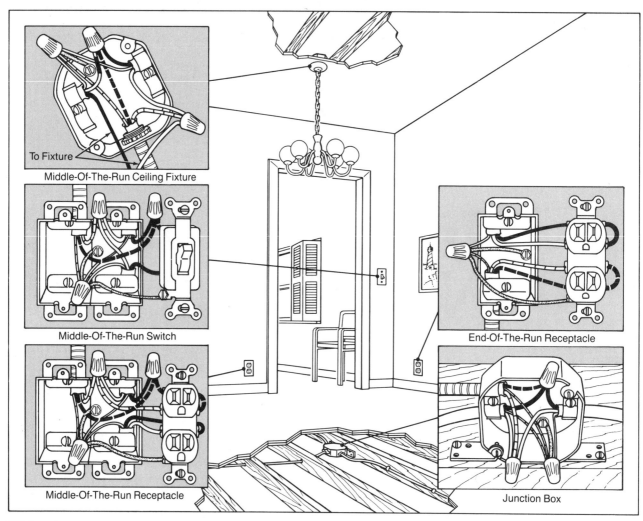

To Fixture
Middle-Of-The-Run Ceiling Fixture

Middle-Of-The-Run Switch

Middle-Of-The-Run Receptacle

End-Of-The-Run Receptacle

Junction Box

Middle-of-the-Run Ceiling Fixture.
Use a ceiling box to extend a circuit. The fixture is used like a junction box (below). Turn the power off, open the box and expose the wiring. You should see black and white fixture wires connected to two separate cables. If you see only one cable entering the box you cannot use it because your new extension would only receive power when the wall switch was on. If you have a switch loop (page 31), you will have one white wire which should be marked black connected to the black wire of the other cable. If unsure about which is your feed cable, use a voltage tester to identify it. To make your extension, cut the white wire that leads to the fixture. Then, with a wire nut, connect the cut white wires to your 'new' white wire. Connect the black-marked white wire with the power source black wire and your 'new' black wire—all with a wire nut. Likewise, attach the grounding wire of the new cable.

Middle-of-the-Run Switch. For this installation, a ganged box was used. The existing box should have two cables entering it—if not, it is not a middle-of-the-run switch and cannot be used. Identify the feed black wire by using a voltage tester. Shut off the power and remove that black wire from the switch. Cut a short black pigtail and attach one end to the switch. Join the other end to the incoming feed wire and to your 'new' black wire—all with a wire nut. Connect the 'new' white wire to the existing white wires with a wire nut. Likewise, attach the grounding wire of the new cable.

Middle-of-the-Run Receptacle. This box will have two cables entering it. In order to accommodate an additional new cable, you will have to extend the size of the box with an extension or by ganging (page 75). Join all black wires, all white wires, and grounding wires to each other as shown here by making pigtails.

End-of-the-Run Receptacle. The easiest location from which to make a circuit extension—but make sure the receptacle is not controlled by a switch. The box will have only one cable entering it. Hook up your new wires by connecting the black wire to the brass-colored terminal screw and the white wire to the silver-colored terminal screw. Attach the new grounding wire with the other grounding wires using a wire nut. This is now a middle-of-the-run receptacle, as power will be passing through it and on to new outlets.

Installing a Junction Box. Junction boxes are commonly used to extend a circuit or to split an incoming power source into two separate circuits. They must always be accessible (not permanently covered) but they must have a cover plate over them to protect the wiring inside. Tap into a circuit by finding available wires, usually in the basement or attic. Turn off the power and cut the wires at the point where you wish to install the box (along a stud or over a joist). Remove knockouts for the existing incoming and outgoing cables plus one for the new circuit. Mount the box and then slip the cables inside. Using wire nuts, connect the three black wires together, the three white wires, and the three bare grounding wires. To ground the box, run a pigtail from the grounding wires to the grounding terminal screw on the box.

WARNING

When doing these wiring projects, first shut off the power and test to make sure it is off. Also follow the safety procedures on page 5.

Running Cable

You've located the existing box from where your run will begin and you've made the opening for your new box(es) (page 76-83). Now you must run the cable before installing the new box or making any wire connections. Read through this section and then try to find a route that involves the least labor and damage to your house. As you can see from the following examples, there are many different routes that you can take, especially if you have an accessible attic, a basement, or a crawlspace.

Everything presented here involves *concealed wiring*, wiring hidden in the walls or other parts of a house. Depending on your situation, as an option, you might consider using *surface wiring*, an easier type to install because it is run on baseboards or on the surfaces of walls and ceilings (page 91). But if you prefer concealed wiring, first read and understand the steps involved (page 53). Next, calculate your material needs and do your shopping (page 119). Also, refresh your memory about the structure of a house (page 52).

Now, a few words of advice. These jobs could be long and frustrating, so it's best to start with small projects and allow plenty of time. Fishing for cable can require patience; or you could run into a firewall and have to begin again with an alternate route. Make sure that you have an assistant available as many of the wiring runs demand two people. If you've never worked with drywall or lath and plaster before, practice cutting a scrap piece or have a handyman-type friend help you. In any case, be careful, wear appropriate personal protection and use the correct tools.

The Importance of Protecting and Anchoring Cable. Because we can't see concealed cable after it is installed, we tend to think of it as inaccessible. Also, we know where it is. But consider that when you sell your home, another owner could indiscriminately pound holes in the walls, ceilings or trim. Unless you've protected the cable, there could be dangerous consequences. That's why the Code states that when cable is close to a concealed surface, such as it is in notched beams or joists (page 61), you must cover and protect it with a steel plate.

When working with cable also take care to anchor it properly. Straps or staples (below) are used for this purpose.

When you first begin your run at a box, staple the cable close to the box and this will prevent you from accidentally pulling it away from the outlet while working. Then, after the entire run is completed, staple or strap the cable as stipulated by the Code. The Code dictates that cable must be anchored at 4½-foot intervals and within 12 inches of every metal box. Cable must be fastened within 8 inches of nonmetallic boxes if connectors or clamps are not used.

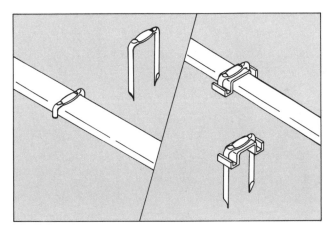

Various methods for anchoring cable include using staples (L), or cable straps (R). Specifically designed for this use, they can be purchased at hardware or electrical supply stores.

RUNNING CABLE PERPENDICULAR TO JOISTS

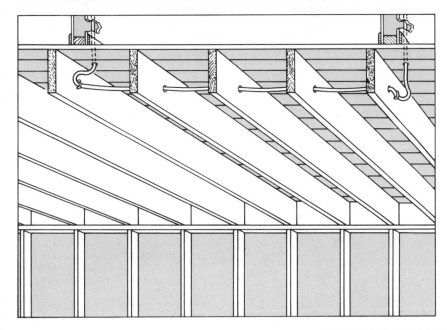

If cable is run perpendicular to joists in a basement or attic, you must drill ¾-inch holes through the center of each joist and pull the cable through. The cable is automatically anchored so you don't need to use staples or straps, except at the ends.

RUNNING CABLE ALONG BASEMENT JOISTS

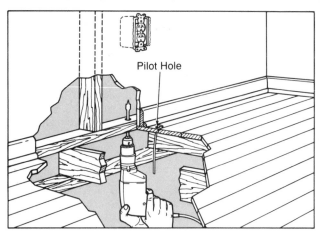

1 Into the floor below the outlet you wish to tap, drill a pilot hole measuring 1/16 inch. Push a thin piece of wire through it so that you can locate it in the basement. Directly in line with the pilot hole, about 2½ to 3 inches from it, drill a ¾-inch hole into the sole plate with a spade bit.

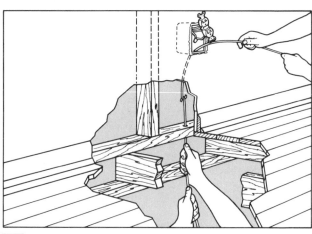

2 With the power off, unscrew the receptacle from the wall and pull it out of the way leaving the wires connected. Remove a knockout from the bottom of the box (page 20). Push a fish tape or a bent coat hanger through this knockout hole and into the wall cavity. Have an assistant push a fish tape up through the hole in the sole plate. Manipulate both of the fish tapes until their ends hook together.

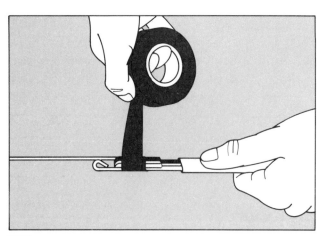

3 Pull the hooked fish tapes down to the basement. Disconnect them. Strip 3 inches of insulation from your premeasured cable and strip the insulation from the exposed wires. Loop these wire ends around the loop in the 'upstairs' fish tape and fold them back securely. With electrician's tape wrap the cable and fish tape tightly so that there are no snags.

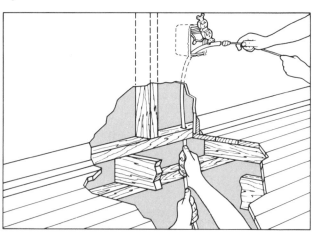

4 Upstairs, pull the fish tape back through the knockout hole. Have your assistant push the cable to make this step easier. When the cable emerges, strip 8 inches of insulation from its end and then fasten it to the box with a clamp (page 20).

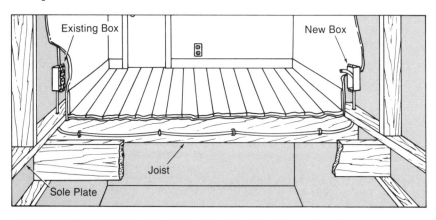

5 Anchor the cable to the joist with a strap or staple and then run the cable along the ceiling of the basement, attaching it to the parallel joist, as shown, with staples. Use caution not to damage the cable when inserting the staples. When you reach the hole you have created for your new outlet, mount a new box (page 75). Then repeat the pilot-hole process (step 1, above).

RUNNING CABLE TO AN ATTIC

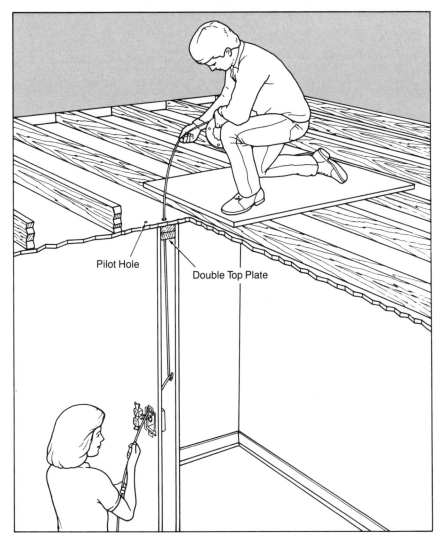

Pilot Hole

Double Top Plate

The procedure is basically the same for running cable to an attic as it is for running cable along basement joists (left). The structural wooden parts of the house have different names but for wiring purposes they function the same. (The floor joists in the basement are like the ceiling joists in the attic; the sole plate is like the top plate). So to begin you will drill the 1/16-inch pilot hole through the ceiling directly above the existing outlet (left, step 1) and then drill the 3/4-inch hole down through the top plate next to the pilot hole (left, step 2). Remove the top knockout from the existing box and push a fish tape or bent coat hanger through. From the attic above, have an assistant drop a fish tape through the hole and after the two are hooked, pull the attic tape down and through the box. Attach the new cable to it (left, step 3). Run your cable to a new box using the same process—usually this will be across the floor joists of the attic, through a hole in the top plate and down a wall to a new opening.

WARNING

When doing these wiring projects, first shut off the power and test to make sure it is off. Also follow the safety procedures on page 5.

RUNNING CABLE BEHIND A PLASTERBOARD WALL

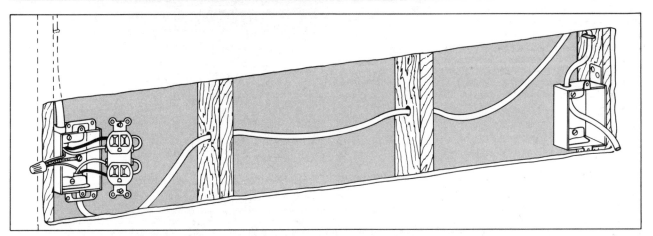

To run cable along a wall that is covered with plasterboard, begin by turning off the power to the circuit and putting on eye protection. With a screwdriver, remove the wall plate and also remove the screws that hold the receptacle in the box. Punch out the bottom knockout for inserting the new cable. From the center of the stud that holds the existing receptacle, to the center of the stud that will hold your new box, measure for a 6-inch wide opening. Very carefully cut out this section of wallboard and save it as a pattern for making a replacement piece. Drill 3/4-inch holes through the studs and run the cable through them. Staple the cable near the box if necessary. Connect the wiring of the new cable to the receptacles. If the installed box has an end-of-the-run receptacle, you should now repair the wall (page 82).

RUNNING CABLE TO A CEILING FOR A NEW LIGHT AND SWITCH

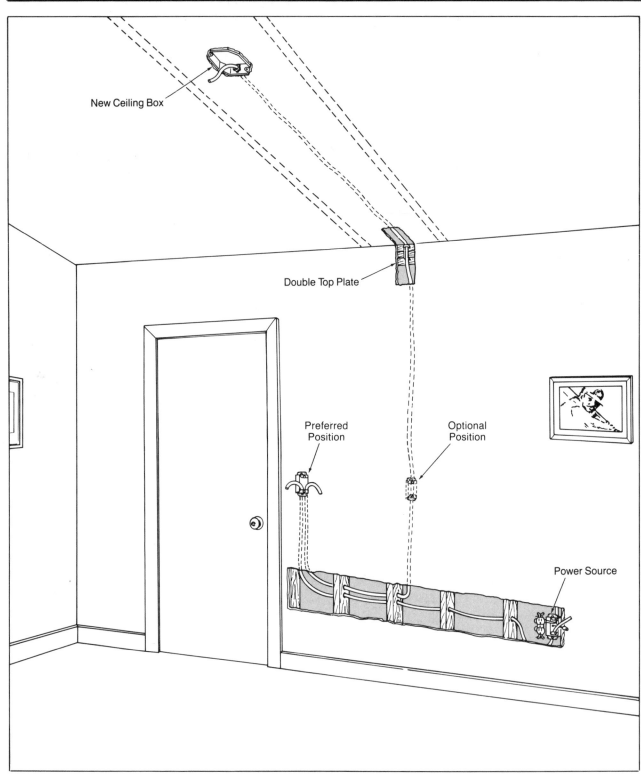

New Ceiling Box

Double Top Plate

Preferred
Position

Optional
Position

Power Source

1 When you have no access to a ceiling from an attic or you wish to extend a circuit from a wall receptacle, use this method. First, locate the direction in which the ceiling joists are running by lightly tapping with a hammer. Position your ceiling box either at one or between two joists (page 79). Follow the path that the cable will run to the wall and, making sure that you're still lined up with the ceiling box, cut a hole approximately 3 inches wide, as shown, both into the wall and into

the ceiling. Use a utility knife, a ball-peen hammer, and a cold chisel. Wear eye protection. If you need to cut lath, use a keyhole or similar saw. Make sure that the wall hole extends below the double plate. Decide on the position of the wall switch. Most people prefer switches by doors but an optional, easier arrangement in this case would be to place it directly below your wall/ceiling hole. After the power has been disconnected, run cable from the power source (receptacle) to

the switch box either using a wall or baseboard route (pages 59 and 62). Feed a fish tape down from the switch box to the cable, attach it, and pull it through the switch hole. Leave about 12 inches of excess cable for your connections. If your switch is not in line with the ceiling fixture, start a second cable from the box and run it to a point below the wall/ceiling hole.

2 Insert one end of a fish tape through the knockout of the switch box or the access hole (whichever is now directly below the wall/ceiling hole). Have an assistant use a fish tape or bent coat hanger to retrieve it. Unhook the two tapes, attach (as shown on page 58) your premeasured cable to the longer tape running upwards, and pull cable through to your switch box. Cut cable leaving about 12 inches for connections.

WARNING

When doing these wiring projects, first shut off the power and test to make sure it is off. Also follow the safety procedures on page 5.

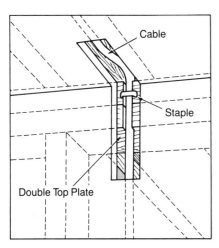

Cable

Staple

Double Top Plate

3 Have an assistant run a fish tape from the ceiling fixture hole to the wall/ceiling hole. After retrieving it, attach the end of the cable and let the assistant pull it through to the ceiling box opening. Allow about 12 inches of cable for the connections. Next, make a notch for the cable at the double top plate of the wall/ceiling hole; use a keyhole saw and a chisel for this step. Make it deep enough so that it will be concealed when the wall is repaired and then staple the cable in place. The Code specifies that you must cover this notch with a 1/16-inch steel plate. Clamp cables in all of the boxes and proceed with the wiring and repair work.

RUNNING CABLE BEHIND A BASEBOARD

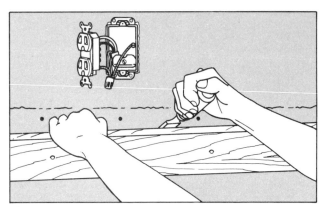 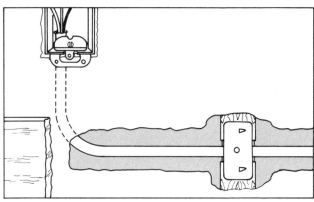

1 Begin by making an opening in the wall where you are going to put your new box (page 76-78). Protect your eyes by wearing safety glasses or goggles. Next, remove the baseboard; for this you will need a wood chisel, a mallet, and a stiff putty knife. If there is a shoe mold present, it should be removed first—by the same basic method outlined here. With the edge of the putty knife, remove the layer of paint that binds the top of the baseboard to the wall. Slip the blade of the chisel at the top of the baseboard and tap it gently with the mallet. Remove it, move several inches along the baseboard and repeat. Do this again and again until the baseboard gradually pulls away from the wall (pounding or prying very hard could break or damage your baseboard).

2 Turn the circuit power off and remove the receptacle from the existing box, with the wires left intact. Punch out the knockout in the bottom of the box. In a lath and plaster wall, chisel a channel from the existing box to the new box. Make ¾-inch holes in the lath at each end, below the boxes. In plasterboard, remove a 2-inch wide section of the plasterboard from box to box. Fish cable up to the existing box, run the cable in the channel and fish it up to the new box. Staple cable to studs as necessary. (Note: Methods vary in regards to studs. You may drill holes through studs for routing cable, make notches in the studs which you should later cover with a ¹⁄₁₆-inch metal plate, or simply staple cable to the studs. Although the last method is widely chosen because it is easier, it is not as safe—the metal plate covering your work assures that driven nails will not damage the cable.) Clamp cable to both boxes, install new box and connect wiring. Replace the baseboard.

RUNNING CABLE FOR BACK-TO-BACK BOXES

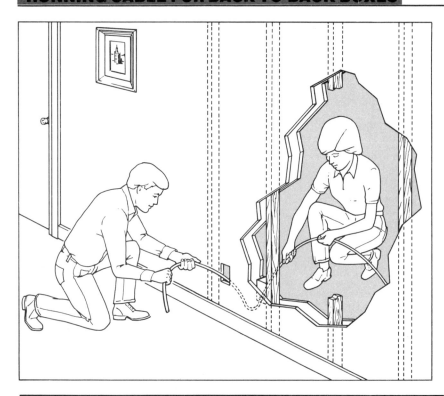

This often-used method of installing wall receptacles is relatively easy to accomplish. Begin by turning off the circuit power and removing the wall plate of the existing receptacle or switch. Pull the device from the box leaving the wiring intact. Now insert a wire or thin coat hanger between the box and wall, into the wall cavity. Search for the wall stud in this manner. Once you have determined which side of the box the stud is on, measure from the edge of the stud to a reference point such as a corner. On the other side of the wall, measure the same distance. Then calculate where the next stud closest to it is; this will typically be a distance of 16 inches center-to-center of studs, but in some newer homes it may be 24 inches. Also, test by lightly tapping on the wall for a solid sound. Cut the opening for the new box next to this second stud—at the same height as the existing box. Remove the knockout in the existing box. Insert the cable through the knockout and push it in the direction of the new box. Have an assistant pull it through. Install the new box, clamp cable in both boxes, and connect all wiring.

RUNNING CABLE TO A CEILING—WITH ACCESS FROM AN ADJACENT ROOM

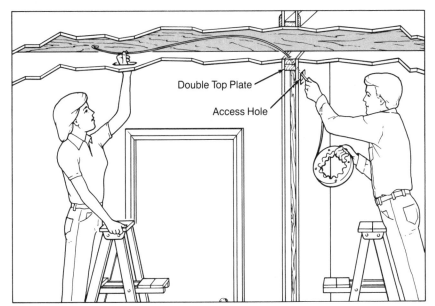

Double Top Plate

Access Hole

1 Sometimes it is easier to work from an adjacent room when extending cable for a ceiling outlet. Begin by making the ceiling hole and marking on the wall the point at which the wall/ceiling hole would be positioned if you were to cut the hole (page 60, step 1). Next, find the point directly opposite in the adjoining room by measuring equal distances from both corners and measuring 4 inches from the ceiling. Drill a small hole in the wall at this point, make sure that you have located the double top plate and then enlarge the hole to accommodate a drill bit. With an extension, drill a ¾-inch hole into the double top plate, as vertical as possible. Insert a fish tape through the hole and push it to the ceiling outlet hole where an assistant will retrieve it. Have the assistant attach cable to it (page 58), and then pull it through the access hole into the adjoining room. Leave about 12 inches of cable at the ceiling hole for connections.

2 Next, insert a fish tape into the wall opening from which you're extending your circuit. Push the fish tape up until your assistant retrieves it with a fish tape or bent coat hanger from the access hole. Have the assistant attach the cable to the fish tape (page 58) and then pull it down and out of the wall opening. Cut the cable, leaving about 12 inches for connections. Install the ceiling box, clamp cable, and continue running cable from the wall opening to the power source.

WARNING

When doing these wiring projects, first shut off the power and test to make sure it is off. Also follow the safety procedures on page 5.

RUNNING CABLE AROUND DOORWAYS

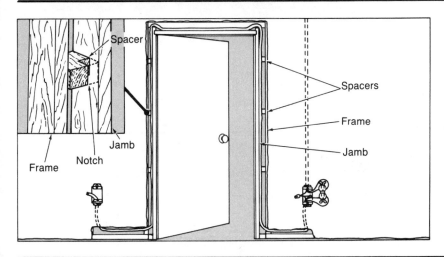

Spacer

Spacers

Frame

Jamb

Jamb

Notch

Frame

To run cable around a door frame, begin by carefully removing the baseboards (left, top) and using the same method to remove the door trim or casing. Run the cable up between the door frame and the door jamb, notching out plaster where necessary. There will probably also be wooden spacers here; make notches in them for the cable run. The Code specifies that you must cover these notches with 1/16-inch steel plates. Clamp cable to the new box, make connections, test, and replace door trim and baseboards.

Installing New Circuits for Lights and Switches

What's shown here may at first look similar to the materials in the Wiring Repairs chapter. But these diagrams are presented with a different approach. This is to aid you in installing circuits or circuit extensions for new switches and light fixtures.

First are the structural considerations. Cable is most easily run across or through exposed attic joists—so that is the preferred route. Cable can also be snaked through walls and behind baseboards as needed for runs from power sources and to and from switches. When diagramming a run, another controlling factor is the power source. As you have already learned, you cannot tap into any kind of outlet. You must choose one of the five types shown on page 56. Once you have decided on it, you can begin your map-making to find the easiest route to take, labor-wise and damage-wise for your installation.

Ceiling fixtures are usually installed in the center of a room but, depending on your decorative needs, you might want to position your fixture off-center. Likewise, the typical location for a switch is near the knob side of a doorway but, if you already have a light source in the room or if you wish to hide the switch, you may choose to position it elsewhere.

Shown here are the two basic kinds of switch-light configurations— a middle-of-the-run and a switch loop. Two-conductor cable is used throughout both of the circuit runs. Then, to give you an idea of a more elaborate wiring design, a 'combination' circuit is shown. For this, three-conductor cable is needed in part of the run. All of the wire connections in the boxes are shown for the switches, light fixtures and, in the last example, the receptacle.

INSTALLING NEW SWITCH-CONTROLLED CIRCUITS

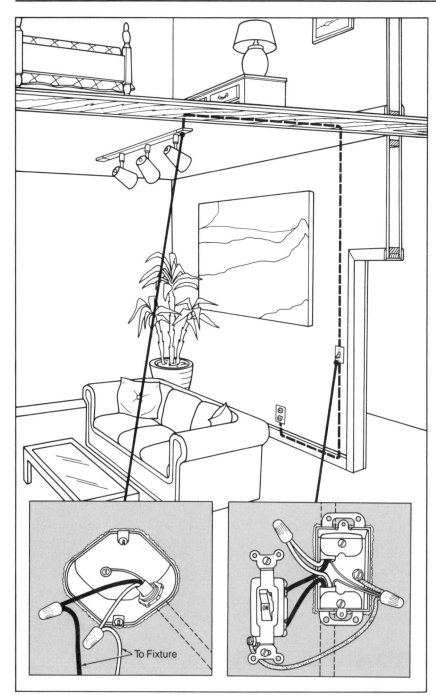

To Fixture

A Middle-of-the-Run Switch. The power source for this wiring run is an existing wall receptacle. Two-conductor cable is used throughout. The single-pole switch wiring is middle-of-the-run and the light fixture is end-of-the-run (page 31). Disconnect the power at the switch, connect the black wire of the incoming cable to one terminal and the black ongoing wire to the other terminal. Join the white wires with a wire nut. Make a pigtail from the green terminal of the switch; connect this and a grounding wire from the box screw—all with a wire nut.

At the light fixture, connect the black wire to the black fixture wire, the white wire to the white fixture, and the grounding wire to the box.

WARNING

When doing these wiring projects, first shut off the power and test to make sure it is off. Also follow the safety procedures on page 5.

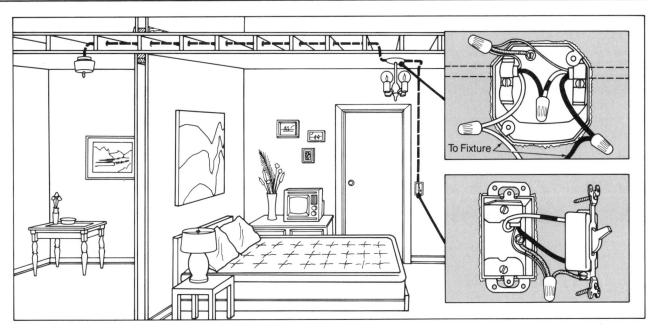

A Switch Loop. The power source for this wiring run is an existing light fixture. Two-conductor cable is used throughout. The light fixture wiring is middle-of-the-run, and the single-pole switch wiring is a switch loop (page 31). Disconnect the power at the new

light fixture, connect the power source white wire to the white fixture wire. Connect the power source black wire to the ongoing white wire. Mark the white wire as black since it will be a hot wire. Connect the black fixture wire to the ongoing black wire.

At the switch, connect the white-marked-black wire to one terminal and the black wire to the other. Attach the bare grounding wire to the box with a $^{10}/_{32}$ machine screw.

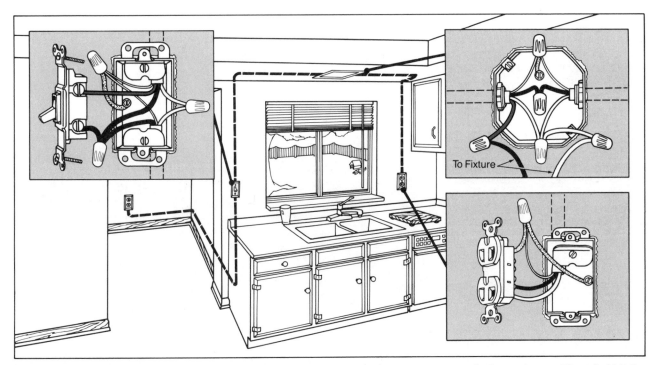

A Combination Wiring Run. The power source for this wiring run is an existing wall receptacle from another room. The first new outlet is a middle-of-the-run, single-pole switch. Next, current will flow to a new middle-of-the-run lighting fixture and lastly it will go to an unswitched receptacle. In order to provide the hot wire for the receptacle and the switch controlled one for the light, three-conductor cable must be used from the switch to the light.

To wire the switch, disconnect the power and connect the black wire of the incoming cable with a pigtail to the switch and the black wire of the outgoing three-conductor cable. Connect the red wire to the switch terminal; the white wires to each other; and the grounding wires to each other. With pigtails, connect the grounding wires to the switch terminal and the box. At the lighting outlet, connect the white fixture wire to the two white wires; connect the incoming and outgoing black wires

to each other; and connect the red wire to the lighting fixture's black wire. Attach grounding wires to a wire nut and then to the terminal screw of the box. At the receptacle box, connect the black wire to the brass-colored terminal screw on the switch; the white wire to the silver-colored screw, and the grounding wire to the green screw. Also add a pigtail to the grounding terminal screw in the box.

Three-Way Switch Wiring Runs

These drawings are provided to give you an idea of how to install three-way circuits or circuit extensions in your home. Stairways are obvious places for such runs, but a quick check around your house might reveal some other areas that could be made more convenient with the addition of light switches.

First, determine where your power source is; it can be one of the five different types shown on page 56. Next, map out routes and determine the easiest one labor-wise and damage-wise. Typically, if the power source is upstairs, you will run cable to the light first, then the switches. If the power source is from the basement or from a wall receptacle, you will run cable to the switch first, then the light. There are three configurations depending on which comes first, in the middle, and last. They are all shown here.

The wiring is indicated for all of the switches and lights in the new installation. Note that three-conductor cable is used often. Instructions are given for the use of switches that contain a green terminal for the grounding wire. You may elect not to use this kind of switch in which case you will simply use a pigtail to ground the box. Remember to calculate the number of wires to be used in the boxes to determine box size (page 74).

INSTALLING NEW 3-WAY SWITCH CIRCUITS

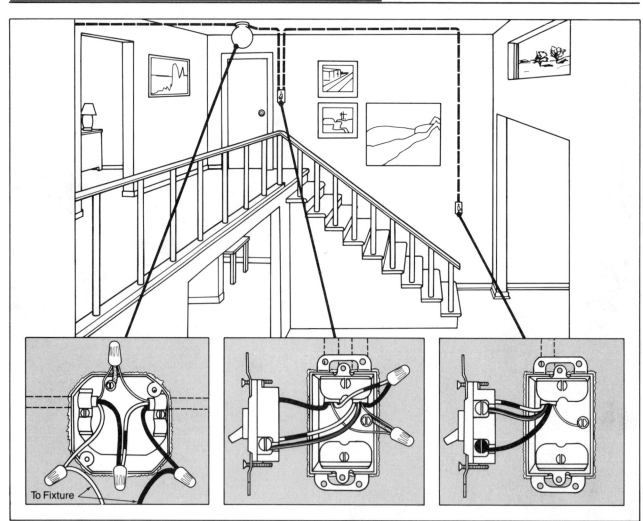

A Light-Switch-Switch Circuit. Note the use of two-conductor cable both from the power source and going out from the lighting fixture. At the first switch, upstairs, three-conductor cable is introduced and run on to the next switch, downstairs.

To wire the ceiling fixture, disconnect the power and connect the incoming power source black wire to the ongoing white wire and mark this white wire as a black wire. Connect the incoming white wire with the white fixture wire. Connect the black fixture wire to the ongoing black wire and connect

the grounding wires, by a pigtail, to the box terminal screw. To wire the upstairs switch, connect the incoming black wire to the common terminal of the switch. Next, connect the incoming white wire to the black wire of the three-conductor cable (be sure to mark it as a black wire). Connect the outgoing white and red wires to the traveler terminals of the switch (again mark the white wire as black). Connect the grounding wires, and run pigtails to the box terminal screw and the green terminal screw on the switch. At the downstairs outlet, connect the black wire to the common

terminal and the red and white wires to the traveler terminals. Connect the grounding wire to the box terminal screw and mark the white wire as black.

WARNING

When doing these wiring projects, first shut off the power and test to make sure it is off. Also follow the safety procedures on page 5.

INSTALLING NEW 3-WAY SWITCH CIRCUITS/CONT'D

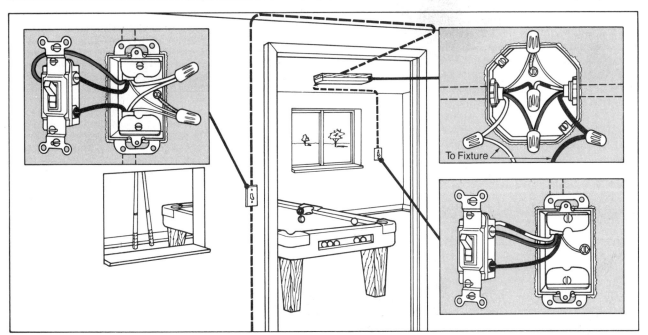

A Switch-Light-Switch Circuit.

Note that two-conductor cable comes into the first box from the power source but three-conductor cable is used throughout the rest of the wiring run.

First, disconnect the power. Then, in the first outlet, the nearest switch, connect the wires as follows: the incoming black wire to the common terminal, the ongoing black and

red wires to the traveler terminals, the white wire to white wire. Ground the box by using a pigtail to the box terminal screw and add another pigtail to the green screw on the switch. In the light fixture box, connect the incoming white wire to the white wire of the fixture. Connect the two red wires and connect the incoming black wire to the ongoing white wire (mark it as a black wire). Connect

the ongoing black wire to the black wire of the fixture. Connect the grounding wires and run a pigtail to the box terminal screw.

At the next switch, connect the black wire to the common terminal, the black-marked white and red wires to the traveler terminals, and the grounding wire, by pigtails, to the green terminal screw on the switch and the box terminal screw.

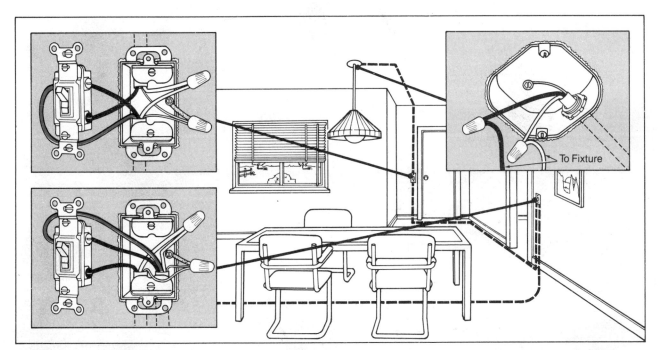

A Switch-Switch-Light Circuit.

Before starting, disconnect the power. Two-conductor cable is used from the power source that runs under the room and to the first wall switch on the circuit. From there, three-conductor is run on to the next switch. At the second switch, three-conductor is introduced into the box to complete the run to the overhead lighting fixture.

At the first switch, connect the two white wires with a wire nut. Connect the incoming black wire to the common terminal of the switch. Connect the ongoing black and red wires to the traveler terminals; connect the grounding wires together with a wire nut and make pigtails to the green terminal screw of the switch and the box terminal screw. At the next switch, connect the incoming red and

black wires to the traveler terminals. Next, connect the white wires and fasten the on-going black wire to the common terminal. Connect the grounding wires with a wire nut and make pigtails going to the green terminal screw and the box terminal screw. At the lighting fixture, connect the black and white wires to the fixture and the grounding wire to the box terminal screw.

Enhancing a Room with Built-In Fluorescent Lighting

Although fluorescent lamps serve well for utilitarian lighting, the bare tubes are not very pleasing to the eye. And yet, concealed behind covers such as cornices, valances, and soffits, they provide a fine, even glow that can highlight draperies or a favorite wall. Such constructions can direct light upward or downward according to your preference and thereby add interesting effects in a room. The color and number of bulbs used are additional factors to consider for varying results.

These concealing structures, often several boards mounted together, are not difficult to build. You only need some basic woodworking skills and tools. Install cable for a switch at the most convenient location (page 64). The cable should be run close to your fixture installation; putting in a box is unnecessary since the channel serves this purpose. Be sure to use rapid-start fixtures (page 47) and to paint the inside of all installations with flat white paint for reflective value.

IMPORTANT

For all of these installations (cornice, valance, soffit, and cove), make sure that you are using the proper hardware for secure attachments to walls and ceilings. Use wood screws when attaching to studs or beams. Use screws with anchors when attaching to plaster, drywall, or brick.

Fluorescent fixtures built into a cornice can give a room warm, even and often dramatic lighting effects.

INSTALLING VALANCE LIGHTING

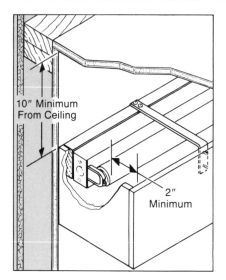

10" Minimum From Ceiling

2" Minimum

Very similar to cornice lighting, this design is different in that it directs light both upward and downward. Valances are often installed over draperies and planning for fixtures within them is a wise idea if you want to highlight the fabric as well as the entire wall. They should be installed no closer than 10 inches from the ceiling. Prepare a mounting block, attach the fixtures, and hook up the wiring in the same way as for cornice lighting (right). The valance board should be at least 2 inches from the fluorescent tube. Mount the board on the inside with screws and angle irons. Construct end pieces so that the lighting fixtures will not show.

INSTALLING SOFFIT LIGHTING

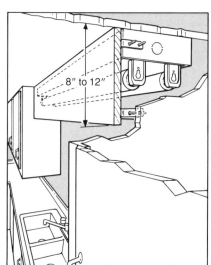

8" to 12"

Soffit lights mounted to the ceiling are very useful over work areas where good illumination is desired. Depending on the depth of the recess, you can install more than one row of fluorescent tubes. If the recess measures from 18 to 24 inches deep, you can install three rows of lamps; if it is only 18 inches, you can install two rows. Mount the channel(s) to the ceiling without using a mounting block. Connect the wiring in the same way as for cornice lighting (right). Attach the soffit board, 8 to 12 inches wide, to the sides of the cabinets with angle irons.

WARNING

When doing these wiring projects, first shut off the power and test to make sure it is off. Also follow the safety procedures on page 5.

INSTALLING COVE LIGHTING

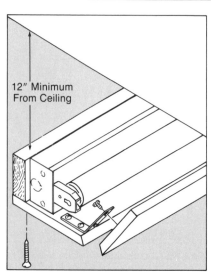

12" Minimum From Ceiling

Cove lighting is used to add general or decorative lighting to a room by reflecting light upward toward the ceiling. The recommended minimum distance from the ceiling is 12 inches. Prepare a mounting block, attach the fixtures, and hook up the wiring in the same way as for cornice lighting (right). Measure and cut a baseboard and a cove board, and bevel the edges of both on the ends where they will be joined. The cove board should be angled outward to diffuse light toward the ceiling. Drill pilot holes in the baseboard, 4 inches apart; secure it to the mounting block with wood screws. Attach the cove board to the baseboard with angle irons angled to the desired degree.

CONSTRUCTING AND INSTALLING A CORNICE LIGHT

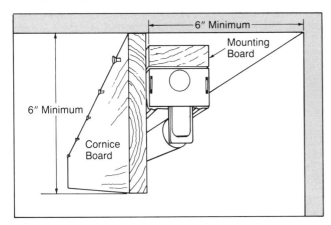

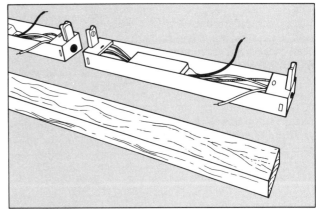

1 Begin by installing a switch for the light and running a cable through the ceiling or wall where you are making your fixture installation. Leave 8 inches of cable for wiring the first fluorescent light. Next, measure for all parts of the cornice. The mounting board will be made out of wood that is 1 inch thick and as wide as the fluorescent light channel (for one light, this will be approximately 3 inches). The cornice board which is screwed to the mounting board, should be at least 6 inches wide and ¾ inch thick. Both the mounting and the cornice board should run the entire length of the wall you intend to light.

2 Cut the mounting board. Punch out the tabs at each end of every fixture except the very last tab where the wiring will end.

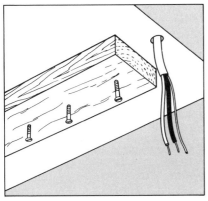

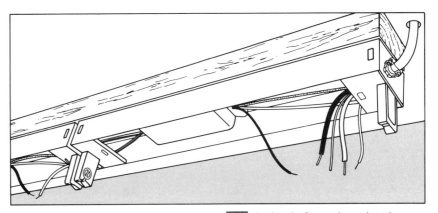

3 In the mounting block, drill pilot holes for the screws; position them 4 inches apart. Draw a line on the ceiling 6 inches away from the wall. Have an assistant help you to hold the mounting block as you screw it into the ceiling.

4 Anchor the fixture channels to the block with screws; butt the ends together. Use a two-part clamp in the end of the first fixture to secure the cable. Connect the fixtures end to end with chase nipples.

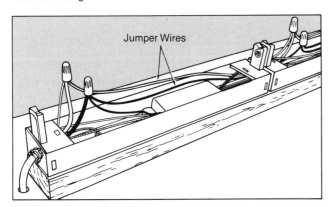

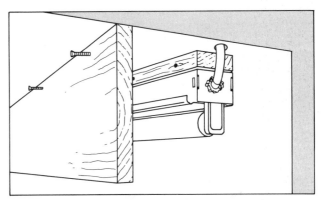

5 Connect the black wire from the cable to the black fixture wire and to a black jumper wire from the next fixture, all with a wire nut. Do the same with the white cable wire, white fixture wire, and white jumper from the next fixture. Connect the bare grounding wire to the fixture terminal screw. Continue to connect wiring in the same manner from fixture to fixture.

6 Drill pilot holes into the cornice board ½-inch from the top and 4-6 inches apart. With an assistant, hold the board up to the mounting block and mark positions for matching pilot holes. Drill these holes and then mount the board with screws. Fill screw holes, touch up or paint over your work as necessary.

Receptacle and Switch Variations

The more 'electricized' your home becomes with new appliances like VCRs and computers, the more outlets you need to keep all of these things running. Once you've calculated the load on a circuit (page 51), made a circuit extension, and run the cable to a wall box, you can now get the maximum reward for your efforts. This is done by installing two receptacles in a box

rather than the standard one. As you can see, the wiring is easy and it can be done with either a 120-volt circuit or three-conductor 240-volt circuit. Be sure to check the capacity of the finished box (page 74) which in this case is actually two boxes ganged together.

Switches Controlling Receptacles. Running cable for ceiling fixtures can be tedious, depending on your access. An easy option to a ceiling light is to install a switch that controls a receptacle where you can plug in a

lamp. Other switch-receptacle combinations can resolve safety or convenience issues. Shown here are directions for wiring a switch to control an entire receptacle or only half of it (a *split-wired* receptacle).

WARNING

When doing these wiring projects, first shut off the power and test to make sure it is off. Also follow the safety procedures on page 5.

INSTALLING A SWITCH AND SPLIT-WIRED RECEPTACLE

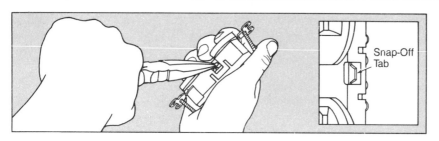

Preparing the Receptacle. To divide a duplex receptacle so that one of them is controlled by the switch and the other remains hot, you must alter it slightly. There is a tab in the center that links the two brass terminals—snap it off with needlenose pliers. Proceed with installation for either a switch loop or a middle-of-the-run switch.

Snap-Off Tab

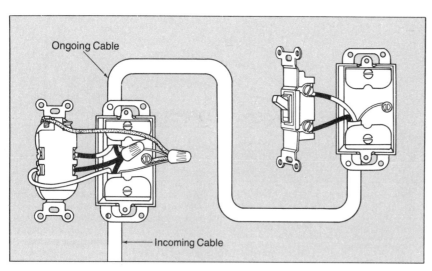

Ongoing Cable

Incoming Cable

With a Switch Loop. A switch loop is identified by the switch being at the end of the circuit (page 31). At the receptacle, connect the incoming white wire to a silver-colored terminal of the altered receptacle. Connect the ongoing white wire to a brass-colored terminal. Mark it as black since it will be a hot wire. Connect a 3-inch black wire to the other brass-colored terminal and connect it to the black cable wires with a wire nut. Connect a 4-inch green pigtail to the green terminal on the receptacle; connect another green pigtail to the box terminal screw; attach all grounding wires with a wire nut.

At the switch, connect the black wire to one switch terminal and the white wire to the other. Mark the white wire as black since it will be a hot wire. Connect the grounding wire to the box terminal screw. Wired like this, the switch controls the top half of the receptacle and the bottom half is always hot.

Ongoing Cable

Incoming Cable

With a Middle-of-the-Run Switch. A middle-of-the-run switch is identified as being between the power source and the receptacle (page 31). For this particular installation, the wiring leading from the switch to the receptacle must be four-wire. At the switch, connect the two white wires. Connect a 3-inch black pigtail to one switch terminal; also connect it to the incoming black wire and the ongoing red wire, all with a wire nut. Connect the ongoing black wire to the other switch terminal. Connect a green pigtail to the box terminal screw and then connect the bare grounding wires to it with a wire nut.

At the receptacle, connect the red and black wires to the brass-colored terminals of the altered receptacle. Connect the white wire to a silver-colored terminal. Attach a 4-inch green pigtail to the green terminal screw, a green wire to the box terminal screw, and, with a wire nut, connect all of the grounding wires. Wired like this, the switch controls the top half of the receptacle and the bottom half is always hot.

INSTALLING TWO RECEPTACLES IN ONE BOX

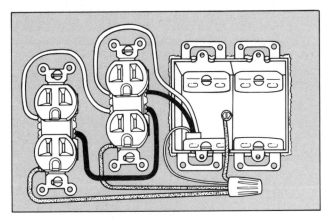

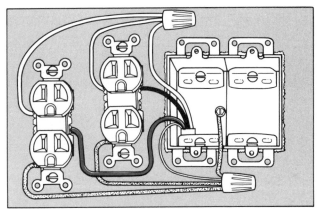

From a 120-Volt Circuit. Gang together two standard outlet boxes (page 75). Run a three-wire cable to the box location and attach the cable to the box (page 20). Install the box and then begin the wiring. Connect the black cable wire to a brass-colored terminal on one of the receptacles and the white cable wire to a silver-colored terminal on the same receptacle. Run a 4-inch black jumper wire from the other brass-colored terminal to a brass-colored terminal of the second receptacle. Do the same with a 4-inch white jumper wire. Attach two 4-inch green pigtails to the green terminal screws, a green wire to the box terminal screw, and, with a wire nut, connect all the grounding wires.

From a 240-Volt Circuit. Gang together two standard outlet boxes (page 75) and connect the 240-volt cable to the box. Run the red wire to a brass-colored screw of one receptacle and run the black wire to a brass-colored screw of the other receptacle. Connect 4-inch pigtails to the silver-colored terminals, and, with a wire nut, connect these and the white wire from the cable. Attach two 4-inch green pigtails to the green terminal screws, a green wire to the box terminal screw, and, with a wire nut, connect all the grounding wires.

INSTALLING A SWITCH AND SWITCH-CONTROLLED RECEPTACLE

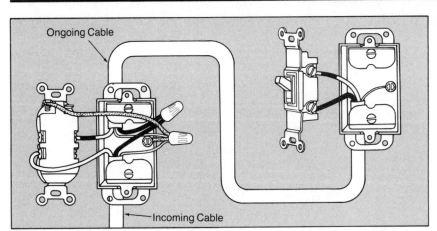

With a Switch Loop. A switch loop is identified by the switch being at the end of the circuit (page 31). Begin by connecting the black incoming cable wire to the black ongoing cable wire. Connect the incoming white wire to one of the silver-colored terminal screws on the receptacle. Connect the ongoing cable white wire to a brass-colored terminal. Mark this wire as black since it will be a hot wire. Attach a green pigtail to the green terminal screw, a green wire to the box terminal screw, and, with a wire nut, connect all the grounding wires.

At the switch, connect the black wire to one of the brass-colored terminal screws and the white wire to the other. Mark the white wire as black since it will be a hot wire. Connect the bare grounding wire to the box terminal screw.

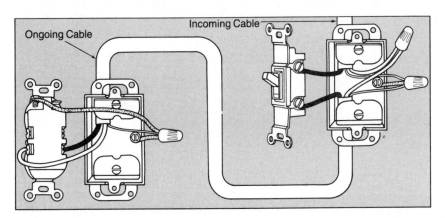

With a Middle-of-the-Run Switch.
A middle-of-the-run switch is identified as being between the power source and the receptacle (page 31). At the switch, connect the white wires of the two cables. Connect the two black wires to the two brass-colored terminals on the switch. Make a green pigtail from the box terminal screw and connect all grounding wires with a wire nut.

At the receptacle, connect the black wire to a brass-colored terminal screw on the receptacle. Connect the white wire to a silver-colored terminal screw. Attach a 4-inch green pigtail to the green terminal screw, a green wire to the box terminal screw, and, with a wire nut, connect all of the grounding wires.

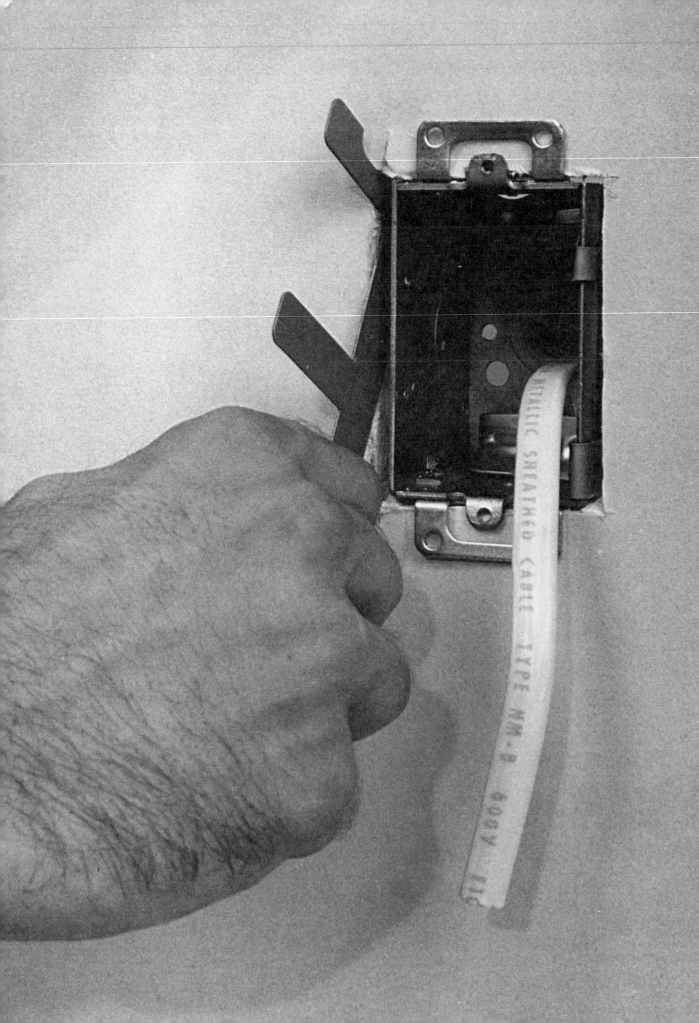

Setting New Boxes into Finished Spaces

Like the last chapter, this one is a 'construction' chapter with no actual wiring in it. But don't despair—you don't have to be a finish carpenter to complete the projects. They might be a little challenging, depending on what kind of wall or ceiling surfaces you're dealing with, but a little care and patience can often make the difference between a sloppy and a neat job.

Remember that in Chapter 6 you were presented with a step-by-step guide to installing new circuits (page 53). This chapter gives you information for steps 1, 9, and 10 of that guide. The very first step was to determine where your new box would be. Then there were seven steps that guided you through the cable installation. Steps 9 and 10, instructed you to clamp the cable to the new box and install it into the wall or ceiling. Clamping is covered on page 20 and the installation is detailed here.

The basic topic here is boxes and, as skilled electricians know, there is a box for practically every situation—thin walls, cramped spaces, multiple outlets, and so on. This potentially complicated subject has been greatly narrowed down for you, the beginner. Boxes are grouped here according to where they are installed (walls or ceilings), and also what they contain (switches and receptacles or fixtures and junctions). This is done for simplicity's sake. When you do begin to make such installations you'll find that the types shown here are the most popular ones available.

The use of boxes is regulated by the *NEC* and in selecting them you should always choose the largest possible for the job. Be sure to use the chart for calculating box size; even though these 'space regulations' are widely ignored, your wiring might not pass inspection if you choose to ignore them.

Boxes—Important 'Little Houses' for Wire Connections

Poor connections can cause electrical arcs and these in turn could ignite insulation, sawdust, or other building materials in your walls. For this reason, splices and connections must be protected by enclosed metal or nonmetallic boxes. Besides housing connections, another purpose of boxes is to provide for periodic grounding connections. Boxes are manufactured of galvanized steel, thermoplastic or, in some areas, Fiberglas™. As with all electrical equipment and devices, look for the UL mark when purchasing.

Boxes need to be completely enclosed when a wiring job is completed and they must also be accessible. They have knockouts on their sides or backs and various kinds of clamps that accept incoming and ongoing cable. See page 20 for the steps in connecting cable to boxes. Remember also that once you have removed a knockout, you must either insert a cable or cover the hole with a knockout seal. Boxes must be covered on their tops; often wall switch-plates serve this purpose. Some boxes such as those used for junctions, have their own metal covers that will not be visible on a wall.

Choosing and Using Boxes—A Simplification. There are literally hundreds of kinds of boxes that can be used by experienced electricians for unique situations. For your purposes as a home electrician, however, you'll find that there are only a few that are in popular use—and the clerk at your hardware or supply store will readily confirm this.

Terminology will vary. This might be confusing at first but a second reading will give you an adequate understanding of the categories, all the various names, and why they are 'misnamed'. A *device box* is one that is used to enclose a switch or receptacle. (Switches and receptacles are *devices*, that is, they are units of an electrical system intended to carry but not utilize electric energy.) Within this book we will refer to a device box as a 'switch/receptacle' box. Generally, it is a rectangular box which may or may not have the ability to be joined or 'ganged' with another box of the same shape to form a larger one for housing more connections. Usually, this type of box is installed in walls.

For electricians, the other type of box besides a device box is an *outlet box*. This is a misnomer since any interruption in a circuit is, in fact, an outlet. Most commonly, these boxes are described by their function or their shape. They function to house wiring for fixtures or for junctions. The boxes used for junctions range in shape from round to octagonal to square. For our purposes here, we will refer to this second type as 'fixture/junction' boxes. Generally, they are installed in ceilings, although the junction box can be installed below flooring (page 56).

One more arbitrary distinction is made here. There is a rectangular-shaped device box (used for switches or receptacles) that has a special feature. Because it is used with surface wiring and is not set within walls, it has rounded, not sharp, corners. It also has a special cover plate. This box is slightly larger than its rectangular counterpart, and it is called a *handy box*.

Hopefully, you aren't terribly confused and will be able to choose the type of boxes that you need for each wiring job or run. The key factors are: what you are installing inside the box and where you are going to mount the box. Again, for simplification, boxes are divided into two categories—switch/receptacle types (right) and fixture/junction types (page 79), and installation procedures are shown for both types—in walls and ceilings, respectively.

NEC Regulations Regarding Usage. There is a point at which a box becomes overcrowded. This is especially true of remodeling work where additional cable is run from existing boxes. For this reason, the *NEC* stipulates the maximum number of conductors that is permitted to be housed in boxes of given sizes. If you discover that you are going to exceed the maximum, you must either remove the existing box

and replace it with a larger box, add an extension ring to it, or gang it with another box of the same depth.

The chart presented on this page is based on the *National Electrical Code,* although the terminology is edited for clarity. Begin by adding up all conductors in a box where there are no fittings or devices. Next, since most boxes do have these 'extras', you will have to examine the box further and make additional deductions. Follow this list to determine how many 'conductors' actually are in a box.

1 Deduct 1 for each conductor that originates outside of the box and ends inside the box.

2 Deduct 1 for each stud, hickey, or internal cable clamp in a box. If one of each is included, deduct 3. (Though an internal cable clamp has two 'loops', count it as 1.)

3 Deduct 1 for each strap used for light fixture mounting.

4 Deduct 1 for any amount of grounding wires entering a box.

For further clarification, a wire such as a pigtail which does not enter or leave the box is not counted and fixture wires are not counted.

CHOOSING BOX SIZE BY NUMBER OF CONDUCTORS*

TYPE OF BOX	SIZE IN INCHES (H x W x D)	MAXIMUM NUMBER OF CONDUCTORS			
		No. 14	No. 12	No. 10	No. 8
Switch/Receptacle	3 x 2 x 1½	3	3	3	2
	3 x 2 x 2	5	4	4	3
	3 x 2 x 2¼	5	4	4	3
	3 x 2 x 2½	6	5	5	4
	3 x 2 x 2¾	7	6	5	4
	3 x 2 x 3½	9	8	7	6
Fixture/Junction	4 x 1¼ (round or octagonal)	6	5	5	4
	4 x 1½ (round or octagonal)	7	6	6	5
	4 x 2⅛ (round or octagonal)	10	9	8	7
	4 x 1¼ (square)	9	8	7	6
	4 x 1½ (square)	10	9	8	7
	4 x 2⅛ (square)	15	13	12	10
Handy Box	4 x 2⅛ x 1½	5	4	4	3
	4 x 2⅛ x 1⅞	6	5	5	4
	4 x 2⅛ x 2⅛	7	6	5	4

* Use the list on this page to determine how many conductors are in a box.

Installing New Boxes in Walls

As mentioned earlier, boxes have been divided into two categories within this chapter: switch/receptacle boxes and fixture/junction boxes. (Handy boxes will be covered in the chapter on Surface Wiring, page 92.) In home wiring, rectangular switch and receptacle boxes are almost always installed in walls. An exception might be a special floor receptacle, but for the most part walls are the most convenient locations for such devices. As shown on page 52, walls are composed of vertical supports called studs to which boxes are usually mounted.

If you are installing boxes in unfinished walls, the task is simple and a good choice is easy-to-mount nonmetallic boxes which are simply nailed into studs. Metal boxes are also available with fixed or optional brackets. These also can be nailed or screwed into exposed studs. If brackets are attached, use screws, as it is difficult to hammer nails without damaging the box.

When working with finished walls, you will not need to locate wall studs; boxes will be fastened by various methods into the walls. Sometimes the ears of boxes are adjusted and used as fasteners for screwing into the wall surface. Optional brackets also help to secure and support boxes in some walls. Special side clamps anchor boxes into wood walls. When you discover that you aren't using a large enough box, you can either rewire a completely new box or, if the original is a standard metal one, gang it to another box of the same depth.

Walls are composed of various materials—the most common in modern homes being drywall or gypsum wallboard, a plaster-like substance compressed between sheets of paper. Drywall is relatively easy to work with because it cuts easily. Older lath and plaster walls pose more problems because laths must be cut very carefully so as not to jar plaster loose and make the patching job larger. Plaster over metal lath walls are not very prevalent but they too must be handled with special care. Wood paneled or hardboard walls are not complicated and, depending on the thickness of the wall, will accept a special box with side clamps.

Very basic carpentry skills and tools are required for these projects. If you have never used drills or saws before, it is recommended that you practice first, work very slowly, and/or have a skilled friend assist you.

TYPES OF SWITCH/RECEPTACLE BOXES

Standard Metal Box for Ganging. Available in a variety of sizes and shapes, to be used according to the number of conductors in the box (the most commonly used box, shown here, measures 2½ inches deep). Contains small ears on the top and bottom for mounting into plaster and lath walls, and holes in the back sides for nailing into studs or beams. When it must be enlarged, the side may be removed and the box connected, by screws, to another box (ganging). Has a small threaded hole in the rear for attaching the bare grounding wire with a ¹⁰/₃₂ machine screw.

Metal Box with Mounting Bracket. With a bracket or flange on the side for easy mounting, this box is usually nailed into exposed studs. Also contains a screw hole in the back for attaching the ground wire.

Metal Box with Optional Bracket. Sometimes called a 'drywall box', this box with its optional brackets is used in drywall or in plaster-over-metal-lath walls. Also contains a threaded screw hole in the back for attaching the grounding wire.

Metal Box with Optional Side Clamps. Used in wood or wood-paneled walls, this box has clamps that expand behind the wall when the side screws are tightened. The adjustable ears also help to secure it in the wall. Also contains a screw hole in the back for attaching the grounding wire.

Nonmetallic Box. A durable plastic box used in most new installations. Can only be used with nonmetallic sheathed cable. Contains nail holes for mounting to wall framing. Incapable of being ganged like metal boxes —but larger plastic boxes are available for doubling switches or receptacles. Unlike metal boxes, it contains no screw hole in the back for grounding. Instead, the bare grounding wire is attached to the green terminal screw of the receptacle or switch.

SETTING BOXES INTO DRYWALL

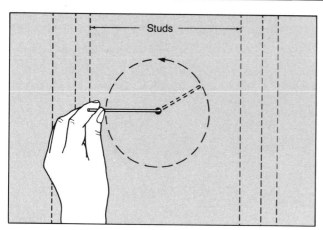

1 Drill a very small hole and insert a piece of bent wire. Check for obstructions by twirling the wire.

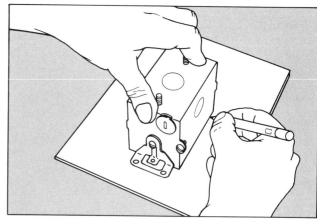

2 Make a template out of thin cardboard by tracing the outline of a standard box, omitting the ears at the top and bottom of the box. Place the template on the wall, centering it over the small test hole and aligning it as level as possible. Trace around it.

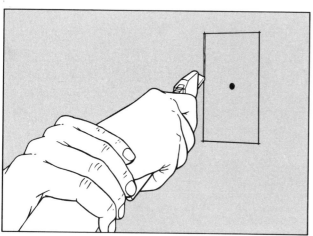

3 Carefully cut out the rectangular piece with a sharp utility knife or a small saw made specifically for cutting drywall, pressing very firmly. If the piece is difficult to remove, use a block of wood and a hammer to push through.

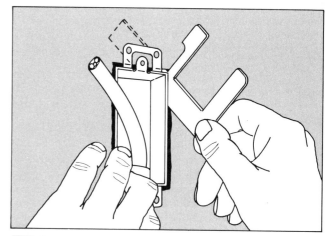

4 Connect the cable to the box (page 20). Slide a side bracket between the box and the wall opening. Put the top in first; then the bottom. Pull an arm of the bracket towards you, gripping tightly.

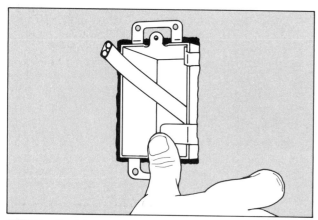

5 Bend both bracket arms inside of the box. Repeat the process, inserting a bracket on the other side of the box.

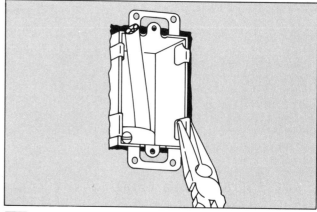

6 Tighten all four of the brackets with a needlenose pliers.

SETTING BOXES IN WOODEN WALLS

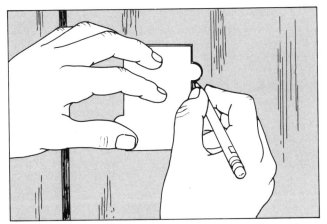

1 Follow steps 1 and 2 for setting boxes in drywall (left), only use a box with side clamps if the wall is ⅜-inch or thinner. If the wooden wall is thicker than ⅜-inch, use a standard box. Trace the outline of the clamps (if the box has clamps) but not the ears. With the template held against the wall, mark the outline.

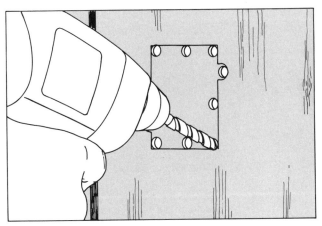

2 Using a ⅜- or ½-inch bit, drill holes in the corners of the outline and also in the spaces for the side clamps (if the box has clamps).

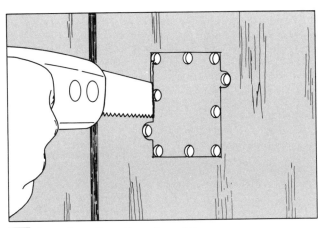

3 Cut on the outline with a sabre or key-hole saw, being careful not to puncture through the other side of the wall.

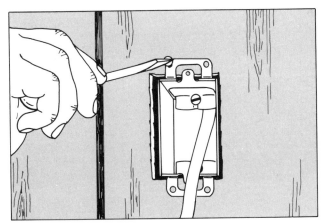

4 Pull the cable through and clamp it to the box (page 20). If the wood is thick, attach the box to the wall using small wood screws through the ears.

5 If the wood is thin, push the box in until the ears are flush with the wall. If necessary, remove the box and adjust the ears. Reinsert the box and, holding it in place, tighten the side clamp screws until taut. The side clamps will bend behind the wall, gripping the box into place.

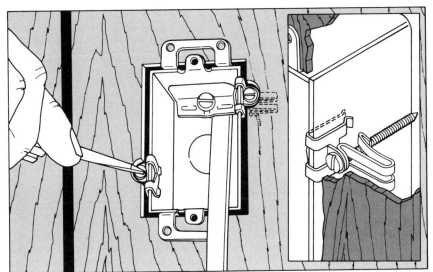

IMPORTANT

When doing these projects, protect your eyes by wearing safety glasses or goggles.

SETTING BOXES INTO LATH AND PLASTER WALLS

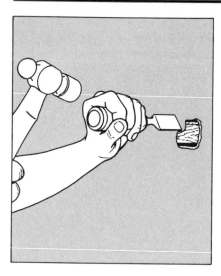

1 Follow step 1 for setting boxes in drywall (page 76). Using a standard box, make a template. Break into the wall at the test hole with a chisel and ball-peen hammer removing just enough plaster to expose one lath.

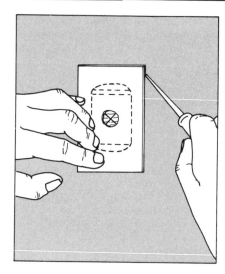

2 Mark an X at the center of the exposed lath. Puncture a small hole in the center of the template. Match the X with the hole and then trace around the template.

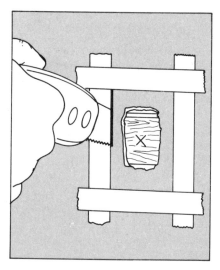

3 Apply masking tape around the edges of the outline to help prevent plaster from loosening. With a utility knife, score the outline. Using a sabre or keyhole saw, very carefully and evenly cut through the plaster and lath.

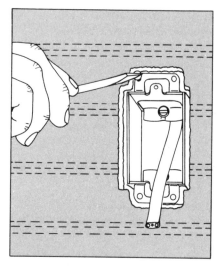

4 Insert the box noting where the ears will be. Remove the box and the masking tape. Chisel a little more plaster away for the ears and mark the positions for screws into the ears. Drill pilot holes for these screws. Connect cable to the box (page 20), put the box into the wall, and tighten the ear screws into the lath.

SETTING BOXES INTO PLASTER-ON-METAL LATH WALLS

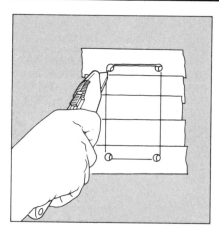

1 Locate a space for the box using step 1 on page 76. Make a template for a standard box (with optional brackets) and then cover the entire area with masking tape. Trace the outline of the box from the template. Drill ⅜-inch holes at each corner and cut on the outline with a utility knife.

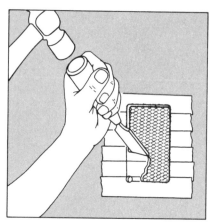

2 With a chisel and ball-peen hammer, carefully chisel the plaster away from the metal lath.

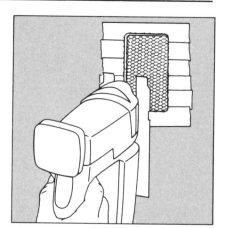

3 Using a hacksaw or sabre saw fitted with metal-cutting blade, insert the blade at one corner and saw very gently. This step must be performed with special care, so that you don't cause damage to the surrounding plaster. Install a box using the optional brackets the same as you would into drywall (page 76, steps 4-6).

Installing New Boxes in Ceilings

More than likely you will be installing a light fixture or a junction box in your ceiling. Boxes for these purposes are either round, octagonal or square, and they may contain as many as four knockouts. Before selecting, make sure that you have counted all of the conductors and are not exceeding the maximum (page 74).

An additional factor in selecting ceiling boxes is access. As shown on page 52, ceilings are composed of horizontally placed boards called joists. Depending on how your home is built, you may or may not have access to these joists from an attic or a similar space. Unlike switch/receptacle boxes which do not have to be attached to wooden studs, ceiling boxes, which often hold heavy fixtures, need to be fastened either directly or indirectly to joists.

The simplest boxes, the flange box and pancake box, can be mounted directly to joists. But for light fixtures that you wish to center in a room or for heavy fixtures such as chandeliers and ceiling fans, you must use either a bar hanger or offset hanger box. Though one is used with access from above and the other below, they both function in the same way. The hangers are anchored to ceiling joists and the boxes can be moved to any point along the hangers for ideal positioning in the ceiling.

Your work will be easy or difficult depending on your access. It is much easier to work crouching or stooping than to reach upward from a standing position on a ladder. If your 'attic' has no flooring, be sure to lay down substantial boards to support your body weight and to provide room for maneuverability.

The most difficult of all these installations is in a lath and plaster ceiling with no access from above. A messy, sometimes hazardous job, it quickly justifies wearing goggles and also demands considerable time and care in patching.

TYPES OF FIXTURE/JUNCTION BOXES

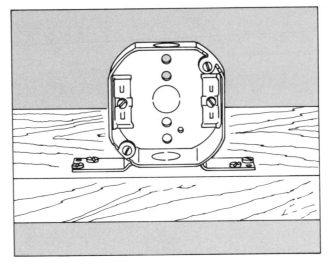

Flange Box. This octagonal box is nailed or screwed into the side of a joist. Although it is easy to install, it has no special feature for positioning it at a specific point such as the center of the ceiling.

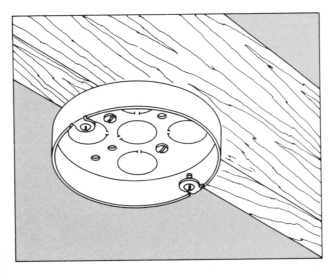

Pancake Box. A round box especially designed for small fixtures weighing less than two pounds. The box will accept only five wires. It is mounted by nailing or screwing it into the bottom of a joist or exposed beam.

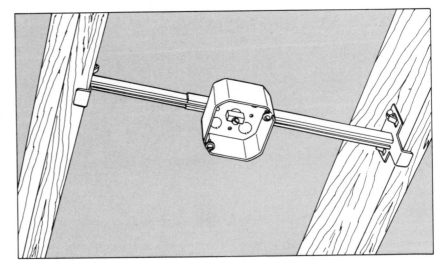

Bar Hanger Box. An octagonal box connected by a movable clamp with setscrew which can be moved back and forth on the bar to a desired position. The bar hanger is a two-piece device which adjusts in length to irregularly spaced ceiling joists. The most popular type to use when you have access above the ceiling.

IMPORTANT

When doing this project, protect your eyes by wearing safety glasses or goggles.

INSTALLING A CEILING BOX FROM ABOVE

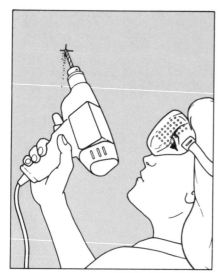

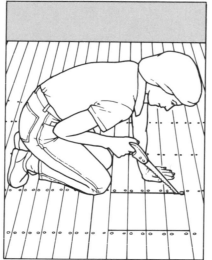

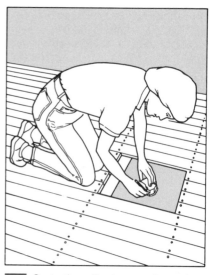

1 If you are working from an attic with no flooring, lay down several boards for the support of your body weight (a loss of balance could send you through the ceiling). If your attic has a floor, you will cut through the flooring as shown here. Begin by drilling a ⅛-inch hole through the ceiling at the spot where you want your fixture. If your ceiling is drywall and you see wood shavings, or if your ceiling is lath and plaster and you meet resistance, you have hit a joist and should move about 3 inches to one side and redrill. Once you have drilled through the ceiling, use a long ¼-inch bit to drill through the upstairs floorboards.

2 Use the drilled hole as a guide and cut out several boards between joists. The joists are located where the floorboards are nailed. Drill ⅜-inch holes on these lines as starting points for sawing. Use a sabre or keyhole saw and save the boards as replacement pieces.

IMPORTANT

When doing these projects, protect your eyes by wearing safety glasses or goggles.

3 Center the ceiling box over the positioning hole. If the box is within 4 inches of a joist, you can simply mount a flange box; if it is further away you should use a bar hanger for the mounting. Regardless, position the box face down and trace its outline on the plasterboard. If the ceiling is lath and plaster, cover the area with masking tape before doing this step.

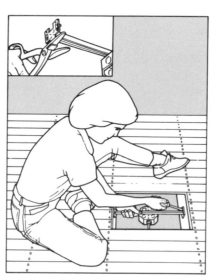

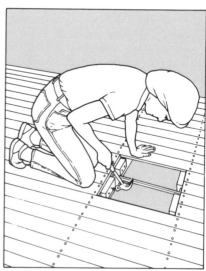

4 With a ⅜-inch bit, drill holes at each corner of the octagon. From under the ceiling, cut from corner to corner with a sabre or keyhole saw. If the ceiling is lath and plaster, apply wide masking tape first to keep the plaster from cracking. For this reason, and also to help you saw straight, you can hold a 1 x 6 board next to the cutting line.

5 Remove the necessary knockout(s) and connect cable(s) to the box. Position the box on the bar hanger using the stud and stud screw; tighten. With metal shears, snip off the tabs at each end of the hanger. Adjust the hanger to fit between the joists; then, using the prongs at the ends of the bar hanger, nail it into place. Drill pilot holes for the screws and screw the hanger into the joists.

A flange box is installed by positioning it from below, drilling the pilot holes, and screwing into the joists.

6 Once you have installed the new box with connected cables, cut two wooden cleats measuring 2 x 4 inches. Nail them to the joists as supports for the floorboards. Replace the boards and nail them to the cleats.

INSTALLING A CEILING BOX FROM BELOW—IN DRYWALL

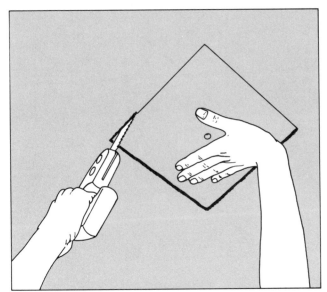

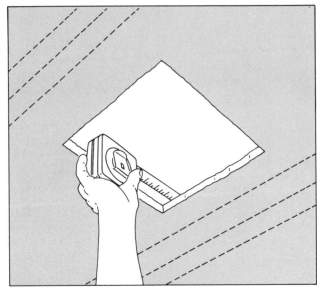

1 Rather than trying to patch a small hole in drywall, it is easier to remove a large piece that extends from joist to joist. Begin by drilling a ⅜-inch hole where you want the box. Insert a piece of bent wire and twirl it to check for obstructions (page 76). With a keyhole saw, cut out a small square about 8 inches wide.

2 To determine where to cut the large piece of drywall, insert a steel tape and measure from the opening to the end where it meets a joist. Do this twice on one side, mark the spot on the ceiling and then add ¾-inch to it. These marks will give you positions in the center of the joists. Repeat the process on the other side of the 8-inch square.

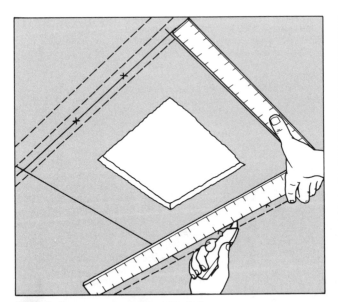

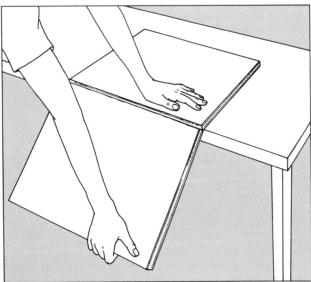

3 Using a steel square, connect the points by drawing lines and then measure to make a 16-inch square with the smaller hole directly in the middle. (Depending on how your home is built, this might be a 24-inch square.) With the steel square and a utility knife, score carefully on all four lines; it isn't necessary to cut entirely through the drywall, just the paper coating.

4 Cut diagonal lines to all four corners and then break up the drywall with a hammer. With a utility knife, cut away the paper coating. Pull any nails from the exposed joists and with a chisel break away any drywall stuck to it.
   Measure for a new piece of drywall; it should be the same dimensions as your hole minus ⅛-inch on each side. Score through the paper coating on one side with a utility knife. Place this edge over a straight board or tabletop; rap sharply on the jutting edge and the drywall should break cleanly at your line. Cut the paper coating on the reverse side. Repeat until you have the correct size piece. Position the ceiling box in the center of it and trace an outline of it.

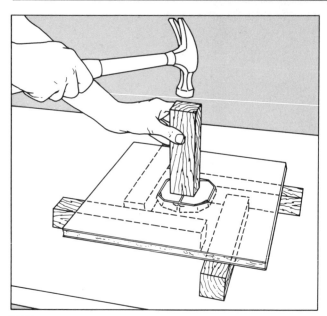

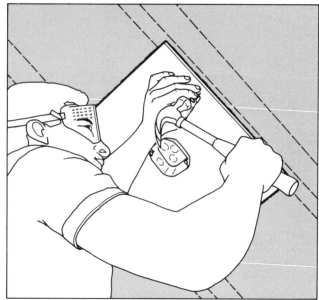

5 Cut through the outline with a utility knife; also cut a line in the middle of the 'box hole'. Elevate the piece of drywall on four pieces of wood with the box outline meeting the edges of the wood. With a small block of wood and hammer break through the drywall; cut away the paper.

Install a bar hanger and ceiling box with cable connected to it (page 80, step 5). Make sure that the box will fit flush with the ceiling once the drywall is repositioned before doing this installation.

6 Either have a helper hold the new dry-wall piece in place or use your forearm to brace it. Drive drywall nails at each corner of the square into the joists, hammering just hard enough to slightly dent the drywall but not break through the paper coating. Continue to secure it to the joists, spacing nails three to four inches apart.

Press drywall joint compound into the cracks with a putty knife or a drywall finishing knife. Apply strips of perforated joint paper—joining, not overlapping the corners. Apply a very thin coat of compound over the strips and feather the edges. Let it dry and repeat three or four times until the paper is entirely covered. Sand the ceiling, paint or finish.

Patching—Completing the Work

Hopefully, you've been very careful in your carpentry work. You might discover that a switchplate fits perfectly to a wall with no visible cracks or holes. On the other hand you might see large gaps that will have to be filled. Gaps around boxes must be filled for another reason other than appearance. If you live in a cold climate, you might notice that outside wall outlets let in frigid air. Stuffing cavities with mineral wool or insulation can help to prevent this seeping effect. In any case, when you do patch around outlet boxes, remember that you must still retain access to the box for additional work.

Patching does not require great skill. Use the correct tools, follow instructions carefully, and always wait until compounds are dry and you should have no problem building a wall surface that is acceptable.

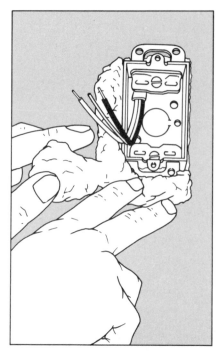

Patching Spaces Around a Box.
Fill gaps around the box with mineral wool or insulation. Next, create a surface around the box that matches the wall surface. If the wall is wood, apply matching plastic wood flush to the surface and sand until smooth.

If the wall is drywall or plaster, apply thick patching plaster or drywall joint compound to the edges and moisten. Beginning here and working inward, apply plaster to a level slightly below the surface of the wall. With the tip of a putty knife, make scores into this wet plaster. After the plaster is dry, fill the grooves with plasterboard joint cement. Smooth the cement a few inches outward from the gaps with a taping knife or comparable broad-bladed tool. Let the patch dry for several days; sand and then paint or finish the wall.

INSTALLING A CEILING BOX FROM BELOW—IN LATH AND PLASTER

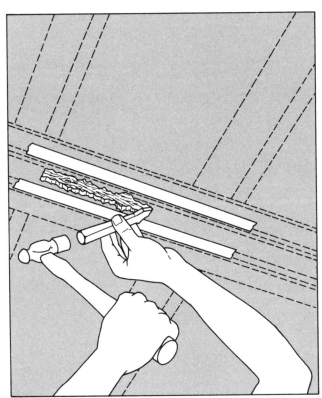

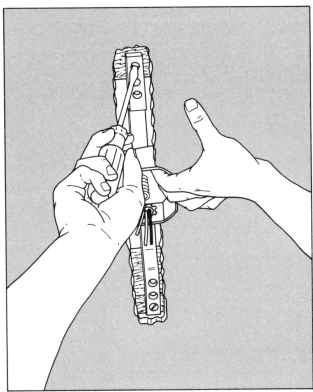

1 A lot of plaster is going to fall on you, so be sure to wear protective goggles. Make a hole where you want to position the ceiling box. With a hammer and cold chisel, chip away enough plaster to find a lath and the direction that it is running. Apply strips of wide masking tape on both sides to help prevent plaster from breaking and then continue to make a channel into the plaster the width of one lath until you locate driven nails. These will indicate the joists at either end.

2 Hold the box to the ceiling and trace an outline of it. Cover the surrounding area with wide masking tape and cut out a hole for the box using the directions on page 80, step 4. With a keyhole saw, cut the lath close to the joists; remove it and the nails attaching it to the joists, being careful not to knock out additional plaster. Drill pilot holes for a bar hanger. Attach the box to the hanger, connect the cable, and then secure the hanger with screws into the joists.

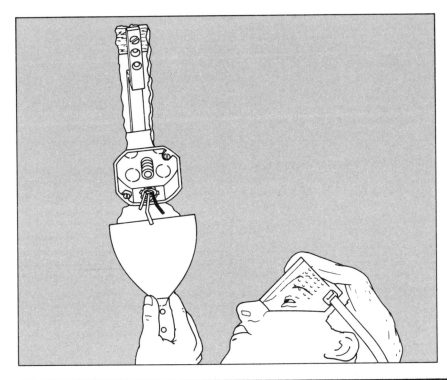

IMPORTANT

When doing these projects, protect your eyes by wearing safety glasses or goggles.

3 Fill the channel with patching plaster about ⅛ inch below the surface of the ceiling. After the plaster has set, apply a layer of spackling compound. Allow it to dry; sand it and paint or finish.

9

Rewiring an Older Home

Older homes can be quite charming with their architectural details, woodwork, and the like. But their wiring can be woefully inadequate or even dangerous. Most homes built since the 1950's have modern electrical materials and service but in many homes built before then, outdated wiring appears.

Some of the problems are insignificant. Older switches will make a loud snap, wiring runs are installed room to room instead of upstairs to downstairs, and a fuse box is in operation—these can all be lived with. Often, however, older homes have the 'octopus' syndrome—extension cords roaming all over the place. People tend to use them incorrectly and for a long time, as if they were permanent wiring. This coupled with frequently blown fuses can be reason enough for serious updating of the electrical system.

You can update an older home by merely adding more outlets (Chapters 6, 7, and 8), adding surface wiring (Chapter 10), or you can, if needed, gain more service by adding a new main service entrance panel. If your fuse box is full, you can add a subpanel for more branch circuits. Neither of these projects is extremely complicated or difficult but they do require the utmost in caution. Your power will have to be turned off by the utility company at the service meter.

This chapter will tell you what to check for in your older home and it will help you decide what to do about the problems that you find. Whether your structure is a current residence, a rental property, or a vacation home, you should understand the situation. Even if you decide not to fix it yourself, you will know whether or not to call an electrician.

Checking the Old System

When you modernize old wiring there are several important points to keep in mind. Remember that whoever worked on your home might have been dangerously unfamiliar with the practices of the day. Also remember that codes have changed greatly and what was once acceptable might now be considered hazardous.

Here is a checklist of outdated things that you might find in an older home.
- Armored cable
- Knob and tube wiring
- Poor wiring practices
- Wire insulation problems
- Improper grounding
- Insufficient number of outlets
- Two-wire (120-volt) service
- Inadequate fuse panel

If your older home is outmoded in any of these ways, or if you suspect it is, you should educate yourself about each of the topics, see if they constitute a problem, and then consider the solutions.

Armored cable, commonly called 'BX cable' (a trade name), is a fabricated assembly of insulated conductors in a flexible steel or aluminum enclosure. The wires are wrapped in paper and may or may not include a grounding wire. The armor also serves as the equipment grounding conductor. Still available, it is rarely used in modern wiring except in certain jurisdictions where local codes still require it.

Armored Cable

Also called 'BX' cable (a trade name), this type is rarely used in modern installations although some codes still require it. Not to be used in damp locations, armored cable has been replaced in modern times by nonmetallic sheathed cable.

If you find that armored cable is part of your home wiring, it is not necessary to replace it unless it is damaged or used improperly in a damp area. Some armored cable has no grounding wire; the metal armor itself serves as the grounding conductor. If this is the case and you are tapping into such a circuit for an extension you should use new nonmetallic sheathed cable. Simply connecting the grounding wire of the new cable to the back of the box provides grounding since the box is already grounded by the armor.

Knob and Tube Wiring

Electricity in the past included porcelain knobs which were used to anchor wires to the building framework; porcelain tubes were also inserted in bored holes to serve as insulating bushings through which wires passed.

Knob and tube wiring need not be immediately replaced just because it is old. If the wires are still in good condition and the circuits are not overloaded,

Definitely antique, knob and tube wiring was once the norm. The name comes from the porcelain knobs that anchor the wires to the building framework and the porcelain tubes inserted in bored holes for wires to pass through. Knob and tube wiring is adequate in an old structure, as long as the wire insulation is in good condition.

a total reworking of your entire system is unwarranted. In modernizing, however, it is not wise to consider merging new wiring with this form. Instead, start any expansion at the service entrance panel and run modern cable to all new outlets.

Poor Wiring Practices

Color coding of wires and terminals may have been ignored—or followed in some places but not in others. Cable splices might have been made at random along a run without being enclosed in an approved box. If a casual inspection reveals even one of these problems, thoroughly check the whole system for additional errors.

Inept wiring practices are frequently found in older homes and if the job was done by only one person, you're likely to find a succession of problems instead of one mistake. If, for example, you find black and white wires connected in reverse order at some point in a run of wiring, you can suspect more errors. *But, do not correct such errors on the spot.* If wire colors are reversed at a connection it is likely that grounding connections are reversed at some other point beyond the error. Correcting the initial error could easily cause a short circuit unless other errors are also corrected.

In this type of old work it is extremely important to use caution. Never attempt even a minor repair until current is definitely turned off at the main switch. Where improper wiring is a possibility, removing a fuse or shutting down the circuit is not enough.

If you find a wiring error, check the entire circuit on which you found it. Do this by shutting off current and removing cover plates and outlet receptacles. Do not disconnect them unless errors exist. If errors exist, correct them and then replace the receptacles.

A faulty splice made outside of a box can be corrected even though you will not have cables of the proper lengths for arrangement inside a box. One solution is to use two junction boxes (page 74) which will be connected by a piece of new cable. As usual, make this repair with the current shut off. The other obvious solution is to run new wiring.

Wire Insulation Problems

If the insulation on wires at the service entrance panel is cracked or crumbling, there is a fair possibility that the same condition exists along the entire run or in the entire system of wiring, even inside armored cable. To confirm this suspicion, check the condition of wire insulation in outlets along the circuit. *If you find the same deteriorated insulation at all points, the wiring should be replaced.*

Fraying, cracking, or crumbling can be attributed simply to aging of some old forms of insulation, or to overheating of the wires. Overheating can occur when undersized wiring is overloaded through the use of an incorrect, oversized fuse. Such overloading can heat the wires enough to cause a fire within the walls or it can, in some cases, break down the insulation and create the possibility of a fire later on.

Improper Grounding

In very old homes wiring might have been installed before the house had a metal water system. Instead of being grounded to a metal water pipe, the system might be grounded to a driven ground rod. If your wiring system is not grounded through your plumbing, check to see how it is grounded. Remember that even if it is connected to a metal water pipe, this is now considered inadequate because the Code requires a supplemental grounding electrode. (See page 3 for modern grounding requirements.)

Insufficient Number of Outlets

If the existing wiring is in good condition, you can extend general purpose circuits (page 51) from existing boxes (page 56). This will not add greatly to a load and will give you the convenience of more receptacles within rooms, the lack of which is often a problem in older homes.

If you need more outlets for small appliance or individual branch circuits (page 51), and you're sure that your system can handle them after calculating the new load, then you can install new branch circuits for the outlets or install a subpanel. Use the information in Chapters 6, 7, and 8 to plan and make your wiring runs but keep in mind that many older homes were not constructed in the standard way shown on page 52 and be prepared for some obstacles.

More often than not, however, in an older home, either your service will be inadequate or your service panel will not have ample room. In any case, remember that in the future more and more appliances are likely to be needed and upgrading the home's wiring now will only increase its value.

Two-Wire (120-Volt) Service

Perhaps the most common reason for wanting to add new wiring is the lack of adequate service. The old two-wire 120-volt service is simply not enough for our modern-day lives that are filled with high-powered kitchen appliances and tools, self-cleaning ovens, air conditioners, clothes dryers, electric heat, and so on. To update wiring for extra service, you will have to install a brand new service entrance panel and other elements of your service will also have to be installed. Begin by contacting your local utility company for information.

Inadequate Fuse Panel

Fuse panels are outdated yet within the realm of Code wiring. Often, however, a fuse box will be filled—offering no more places to install new circuits. If panelboard terminals have been overloaded for a long time or are in a deteriorated condition then you will probably want to install a new service entrance panel board and box—this time an up-to-date circuit breaker type. Other reasons for doing this could be to gain more service or to simply acquire an easier-to-use circuit breaker type box. Finally, too, there is a financial justification, as fuses will no longer need to be purchased.

Another cure for such a problem and an answer to other problems is to install a subpanel. A subpanel acts like a 'mini service panel' when you need to install considerable wiring in a remote location from your main service entrance. This is accomplished by running a properly sized cable to a separate box; from it two to six branch circuits extend out and supply power.

In an older home, subpanels are used when the fuse box is overloaded and new circuits are needed for workshop appliances or individual appliance circuits. They make even more sense if such appliances are in a remote area of the residence. Instead of running three branch circuits the entire distance, one large cable will suffice.

Preparing for a New Service Entrance Panel

If you need a new service entrance panel you must first determine what your total load is (page 51); then you can purchase your panel. They come in capacities of 100, 125, 150, and 200 amps and though the Code recommends a panel to match your load, it is best to get one of the next size to allow for future installations. In each amperage rating, panels come in varying sizes according to the number of circuit-breaker openings they contain. For the average modernization project, this will be the number of circuits you already have plus at least six more.

The panel and covers are sold separately and you will need to get a panel to suit your particular installation requirements. If you are mounting flush to a wall, the panel will be attached to two studs. The surface mounted type, for masonry walls, is simply screwed to a plywood backing which is secured to the wall.

You will also need to buy the supply wiring which leads into the panel from the service meter. Generally, this will be *service entrance concentric (SEC) cable*, but check with your local code before purchasing. This cable should be of the correct size corresponding to the amperage demand, as shown here.

Service Entrance Cable

Amperage	Copper Cable Size
100	No. 4
125	No. 2
150	No. 1
200	No. 2/0

Variations in the Procedure. Here, you will be shown the most basic and ideal situation—a simple replacement with the most common type of service entrance panel. However, there are other situations you might encounter.

■ Meter socket, weatherhead or underground connection, plus the conduit and wire for these connections. If you're going to receive new, higher amperage service, these components will also have to be replaced. Call your local utility company for assistance and advice.

■ A choice of service entrance panels. There are panels available with split-power buses. Although the straight bus type (shown in the illustration) is more expensive, it is more versatile. There are also panels with built-in main circuit breakers or panels with lugs for installation of a separate main disconnect. There are even boxes that house both the service meter and the panel. Local codes may vary in their recommendations; be sure to check before doing your work.

■ Optional surge arrester. After you have made your installation you might want to add this device. It protects your home from damage when lightening strikes nearby power lines.

■ Troubleshooting for inadequate wiring. The illustration shows the ideal situation of the wires being the correct length but there is a chance that some or all of your branch circuit wires will end up being too short for connections at the panel. Some electrical inspectors will permit one or two splices within a panel box but this is the exception rather than the norm.

It is a better idea to install a junction box (page 74) next to the new service entrance panel, or, if many wires are short, to use the old service entrance box as a junction box. To do this, remove its insides, connect new and old cables with wire nuts, and run new cable through unopened knockouts. Cover this 'junction' box with a solid sheet-metal cover.

The Position. Unless you have a separate main disconnect as mentioned above, your box should be as close as possible to where supply lines from the meter will enter the house. As required by Code, there must be at least three feet of clear working space in front and on each side, and it should not be in a wet or confined location.

Precautions. Since the service panel is the first part of the new service to be installed, you must be sure that the supply of power to the entire house is shut down.

Completing the Project. When you are finished with your part of the work, installing the service entrance panel, you should have the meter installed, the weatherhead or underground service installed, and all the wiring that connects them. Then the electric company will make the final connections to restore service.

REPLACING A SERVICE ENTRANCE PANEL

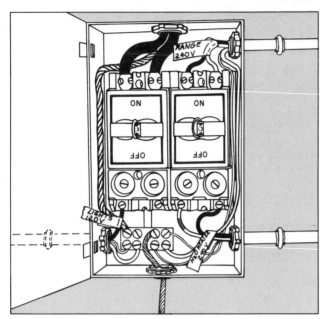

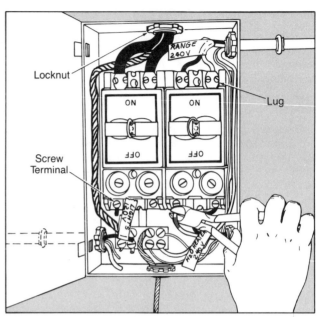

1 With the power cut off to the panel *from the meter*, begin by marking with tape all of the branch cables according to their function. Mark the black hot wire of each 120-volt circuit. Group together and mark both the black and red wires of 240-volt circuits. Any white wires connected to fuses are hot; mark them red as hot wires, trace their corresponding black wires, combine them as a pair, tape and identify.

2 Protect your eyes with safety glasses or goggles. With pliers, sever the wires as close as possible to screw terminals. At the lugs (set-screw pressure connectors), loosen the setscrews and pull out wires. Unscrew lock nuts where cables enter the box and loosen connector screws outside of the box. Remove staples near the box, being careful not to damage the cable sheathing. Unscrew the screws that hold the panel to the wall and with an assistant pull the panel down from the wall while removing all wiring. Prepare the wall surface for the new box either by cutting into the wall for a flush-mounted box or adding ¾-inch plywood with an eight-inch clearance on all sides.

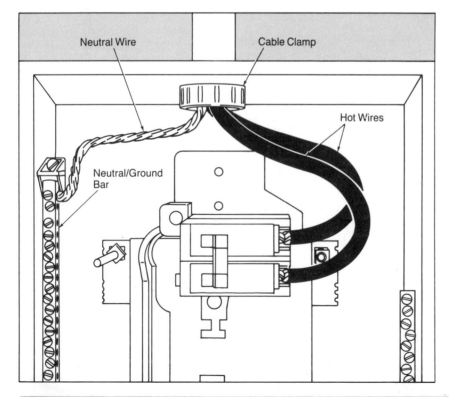

CAUTION

To replace a service panel, begin by shutting down power at the service meter. Called 'pulling the meter' or 'unplugging the meter', this step must be done by a licensed electrician with approval by your local utility company.

5 Connect the service conductors, attaching the hot wires to the main disconnect switch terminals and the bare neutral wire to the neutral/ground bus bar. For the hot wires, strip about one inch of insulation and connect as to any other terminal (page 19). The strands of neutral wire should be twisted together, cut bluntly and secured tightly under their terminal screw. Tighten the cable clamp at the top of the panel. Pull all of the branch circuits into the box, place lock nuts on the cables, strip the sheathing close to the connectors and tighten. Next, make all remaining connections on the new service panel using the instructions on page 22 and the next two illustrations as a guide.

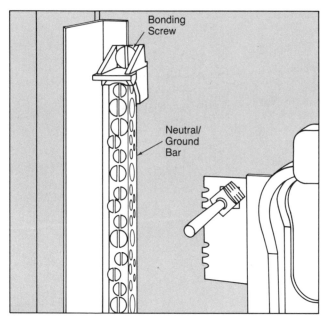

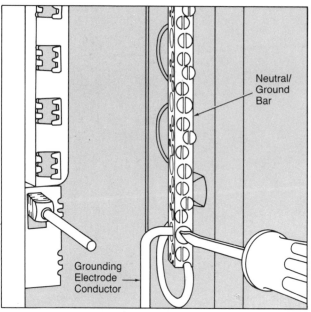

3 Bond the panel box to the neutral/ground bar by tightening the bonding screws located at the top and bottom of the bar (as shown), or, if a strap is provided, screwing one end into the panel, the other end to the bar. Then mount the panel, either screwing a flush-mounted type into studs on both sides or screwing a surface-mounted type into the plywood. Check the box to make sure that it is level before inserting the final screws.

4 Ground the box by attaching a bare copper wire to the lug at the side of the neutral/ground bar and running it through a knockout to your grounding electrode system (page 3). This wire should be as short as possible. If you are using the grounding electrode conductor from your old panel, make sure that it is the correct size for your new service. The Code requires the size of the grounding electrode to be No. 8 for 100 amp, No. 6 for 150 amp and No. 4 for 200 amp service.

WIRING A SERVICE ENTRANCE PANEL

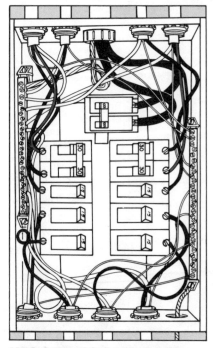

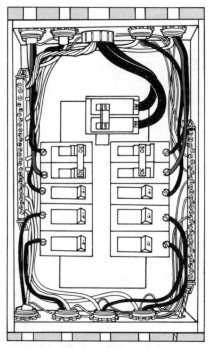

INCORRECT METHOD

CORRECT METHOD

IMPORTANT

Good wiring inside a service entrance panel box includes short, neatly run wires and an orderly arrangement of grounding and neutral wires. Here it's especially important to avoid excessively long and tangled wires.

Adding Surface Wiring

Surface wiring is actually the oldest type of wiring; it was a common procedure in the early days of electric power. Today it is still the easiest way to install branch circuits in homes, although its use is more prevalent in business and industrial settings.

The reason for its simplicity is obvious—circuits are mounted *on the surface* instead of snaked through the walls, floors, and ceilings of a house. Of course the trade-off is the not-s'o-lovely appearance of exposed wiring as contrasted to the finished look of standard wiring, and yet, there are places in the home where it is entirely appropriate.

There are many forms of surface wiring; this chapter instructs you in the installation of two: track lighting and nonmetallic raceway systems. The use of track lighting has simply mushroomed in recent years—many homeowners prefer its special spotlighting effects—and, as you'll see, it's not difficult to mount. Also covered is a nonmetallic raceway system, a quick way to add electricity in recreational or utilitarian areas. For both installations, you should read the manufacturer's how-to instructions along with the material presented here—there might be variations. Also, read and follow the directions in Chapter 6 for extending a circuit.

The process of adding surface wiring has its benefits. If you have little experience or an aversion to doing the construction-type work that regular wiring demands, you might consider it. Also, these are excellent beginner projects that one person can do, without an assistant.

Fashionable and Functional Track Lighting

Track lighting, once seen only in stores and museums, is rapidly gaining in popularity among homeowners. Easy to install, this form of lighting includes decorative 'sliding' fixtures which can be positioned and pointed in varying directions to display artwork or provide illumination over countertops. Generally, tracks are used with spotlights to add a high-tech look to a room although there are also chandelier adapters available.

The track itself is like one long receptacle with current-supplying contacts. Pre-wired, at one end it contains a 'live-end' connector which is connected to the house wiring. Usually available in lengths of 4 or 8 feet, tracks can be hooked up with right-angled, T-shaped, or even X-shaped connectors to create decorative and functional patterns on a ceiling or wall. Regardless of their patterns, all forms of track lighting include a box adapter which is used to tap into an existing fixture box for the power supply.

Purchasing and Mounting. Manufacturers' products and accessories vary so be sure to use only one brand for your entire installation. Before you purchase the system, determine where it will be positioned and connected to the house wiring. Figure out which branch circuit you will tap into and check to make sure that the additional fixtures will not overload the circuit (this is doubtful unless you're installing many fixtures or you already experience frequently blown fuses on the circuit).

Generally, on an 8-foot ceiling, the track should be from 2 to 2½ feet from the wall; on a 9-foot ceiling, approximately 3½ feet. More often than not, ceiling (light fixture) boxes will not be located in these positions, so you will first have to make a run and install a junction box from your main ceiling fixture outlet.

Determine how much wiring and other supplies you need. Depending on the kind of wall or ceiling surface, you will need to purchase expansion shields or masonry anchors for attaching; you should also use insulating washers between all screws in the metal channel.

Fast and Easy Raceways

A raceway installation is the simplest means of adding outlets in a finished structure. It minimizes the need for cutting into walls and permits future wiring changes or expansion with very little work, as compared to in-the-wall wiring. Because raceway installations are visible, they aren't the first choice for rooms in which decor is a priority. And since they can only be installed in dry locations they again have limitations. But for utility rooms, workshops, and recreational spaces they are sometimes the perfect answer; and their ease of installation makes them even more desirable.

Raceways are like huge extension cords but they have these extra benefits: They are much better insulated, and therefore safer. Also, through their wiring, metal coverings or both, they provide grounding which only a heavy-duty extension cord can duplicate. Raceways, furthermore, give you a complete system of wiring; instead of merely receptacles, you have the option of adding boxes for switches and for lighting fixture outlets.

Types of Raceways.   There are many types of raceways available. They vary in composition, installation procedure, and finished appearance and you should consider all these when making your selection. Here are the three most common types:

■ Two-piece channel system. The back is fastened to the wall, wires are laid into place and the front is snapped on. These are available in metal or nonmetallic material.

■ Prewired nonmetallic strip system. Wires are installed before the channel is installed. Receptacles are built into the channel rather than added onto it.

■ One-piece system. Metal or nonmetallic tube-like channels are attached to wall and ceiling surfaces at each end with bushings. Wire is snaked through the channels.

Even within these categories, there are various styles. Some can be mounted directly over a floor baseboard while others replace it. Some have box-size receptacles and others have narrow strip-size receptacles.

The one-piece system is the type shown in the wiring scheme and in the step-by-step instructions on pages 94 and 95. Made of nonmetallic material, it features channels and connectors for easy wiring. This system also accepts receptacles and switches used in regular home wiring although the boxes are different. Boxes, since they are exposed, have rounded corners and fewer 'knock-outs'. Called *handy boxes* or *utility boxes*, they are also a little larger than the norm.

Code Considerations.   The *NEC* has restrictions concerning the use of raceways. Metal and nonmetallic types may be installed in wet locations as long as the proper rain-tight type fittings are used, but all types of raceways should be resistant to impact or physical damage. Nonmetallic raceway must be flame resistant. Most importantly, the size raceway to be used must be determined by the number and size of wires which are run through it.

Most local codes permit homeowners to install their own surface wiring but some do have limitations. As always, it is best to check the local codes in your area before you begin an installation project.

A TYPICAL RACEWAY SYSTEM

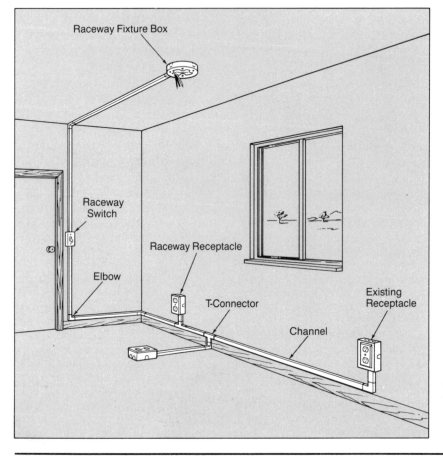

Raceway Fixture Box

Raceway Switch

Elbow

Raceway Receptacle

T-Connector

Channel

Existing Receptacle

Developed for nonresidential buildings, raceways can save you time and trouble in installation. Most often used in recreational and utilitarian areas of the home, their only disadvantage is the appearance of the track and protruding (not built-in) fixtures. In this raceway wiring run, the power source is an existing middle-of-the-run wall receptacle. There are housings for switches and lighting fixtures as well as special connectors for turning corners and branching off in two directions. Before making such a plan, make sure that you have calculated the future load on your branch circuit.

INSTALLING TRACK LIGHTING TO A BOX

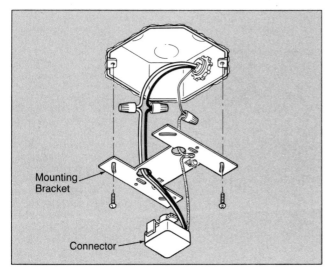

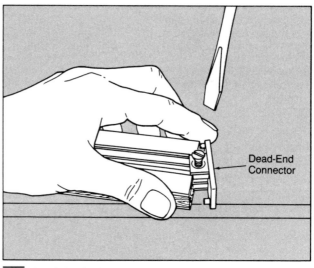

1 First locate or install a fixture or junction box for the connection. Next, run the lead wires from the electrical connector through the provided holes in the metal mounting bracket, as shown. Connect the wires to the box wires using wire nuts—white to white, black to black and green grounding wire to bare grounding wire. Screw the mounting bracket to the box.

2 Attach the dead-end connector to the track and tighten the setscrew. With a ruler or straight edge, mark a line on the ceiling or wall the length of the track.

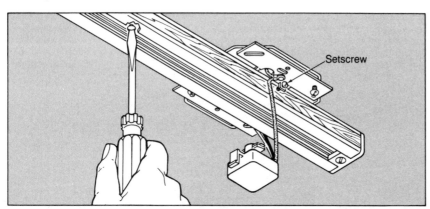

3 Position the track on the mounting bracket. Attach the track to the ceiling using regular screws or screws with anchors, as necessary. Tighten the setscrew to secure the track to the mounting bracket.

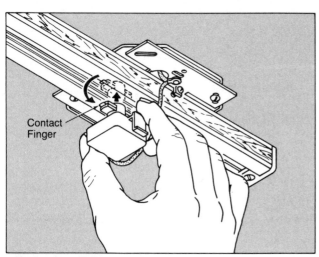

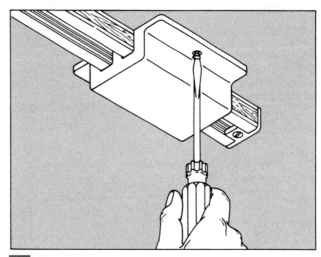

4 Position the electrical connector and twist it 90° into the track. Make sure that the metal contact fingers fit snugly into the two track slots.

5 Place the canopy cover over the mounting bracket, connector, and track and secure it with screws to the mounting bracket.

Planning a Raceway Installation. Follow these steps, in sequence, and then use the instructions on these two pages for installing all of the components of a raceway system.

1 Determine what your source of power is to be (page 56). Remember that not every outlet will be adequate for this purpose.

2 Make a map, complete with measurements, of the proposed raceway system. Considering the appliances and fixtures you intend to use on the branch circuit, calculate what the future load will be (page 51).

3 Purchase all supplies. Wire should be TW type (page 9). Buy black, white, and green. For a 15-amp circuit use No. 14 wire; for a 20-amp circuit use No. 15. Get enough for the entire distance measured plus approximately ⅓ more. Buy all channels, connectors, couplings, and boxes. Channel length should be exactly as measured plus an additional 5 to 8 feet to allow for error. Boxes should be of the same style as the entire system and the size should match up with the number of connections that it will include (page 74).

4 Prepare the existing power source to accept the raceway system (shown above).

5 Install all of the new boxes, for receptacles, switches, and lighting fixtures.

6 Install any necessary extension, elbow connectors or T-connectors. Measure, cut, and install the channels.

7 Connect the wiring and add all covers.

WARNING

Before doing any of these projects, test to make sure that you have turned off the power to the circuit at the service entrance panel (page 4). Also, for eye protection, wear safety glasses or goggles.

INSTALLING RACEWAY COMPONENTS

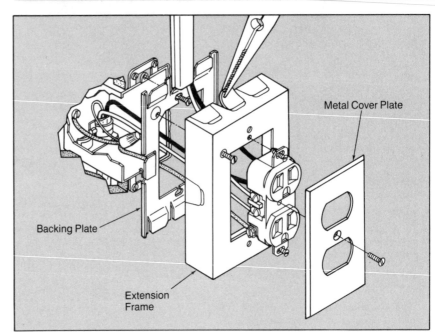

Extending the Circuit—Installing the First Box. The existing box must be extended outward so that the raceway can be connected to it. This is done with a two-part adapter. (Shown here are parts for a rectangular receptacle box; parts and procedures are basically the same for a switch box. If the first box is at a light fixture outlet, the two-part adapter will be round.) With current off, remove the receptacle but do not disconnect the wiring. Protect your eyes by wearing safety glasses or goggles. Slip the backing plate over the receptacle then prepare the extension frame by breaking off, with pliers, the tab where the raceway channel will enter. Put the extension frame over the receptacle, insert the channel into the tongue of the backing plate, and then screw the plate and frame to the box. Connect the wires after the raceway is in place according to the instructions for replacing 120-volt receptacles (page 37).

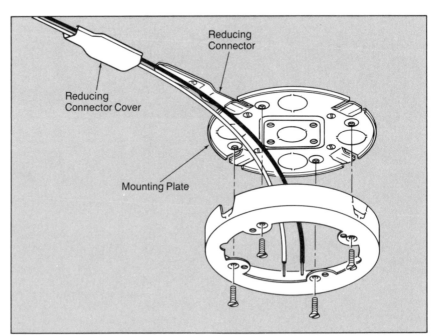

Installing a Raceway Fixture. Install a raceway light fixture box in the same way that you install a switch or receptacle, with one exception: Add a reducing connector between the box and the raceway channel. Mount the reducing connector base so that the larger end of it overlaps the tongue of the mounting plate. Slide the connector cover onto the base. To hook up the raceway channel to the connector, push it onto the tongue. Connect wires using instructions for replacing lighting fixtures (page 44).

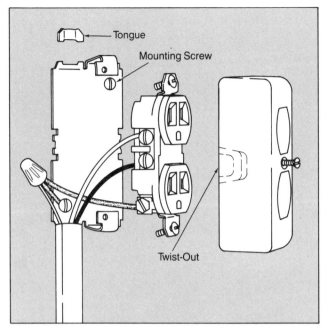

Tongue

Mounting Screw

Twist-Out

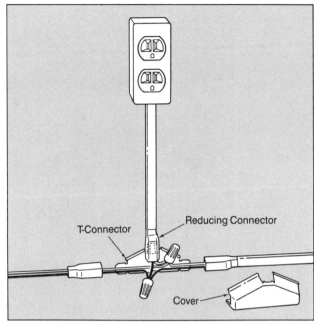

T-Connector

Reducing Connector

Cover

Installing a Raceway Receptacle or Switch.

The components are the same for a receptacle and switch: a metal mounting plate and cover. Begin by removing all of the tongues on the plate except the one that will fit into the raceway channel. Screw the plate into the wall, using anchors in plaster or drywall. Measure for the raceway channels, cut them and install. Next connect wires according to the standard procedures for receptacles and switches (pages 30 and 37) with the exception of the grounding wires. Ground the box by attaching a pigtail to the green terminal screw of the receptacle or switch and connecting it to the screw holding the box to the wall. Mount the receptacle or switch by screwing the ends into the arms of the mounting plate. Remove the twist-out from the cover where the raceway channel will be inserted, place over the mounting plate and screw into the receptacle or switch.

Installing a Raceway T-Connector Junction Box.

When a receptacle is in the middle of a run, as shown in the drawing of a raceway system (page 92), you must install a 'junction box' in the form of a T-connector to hold the six wires and two wire nuts. Reducing connectors are also required to lead the wires into the channels.

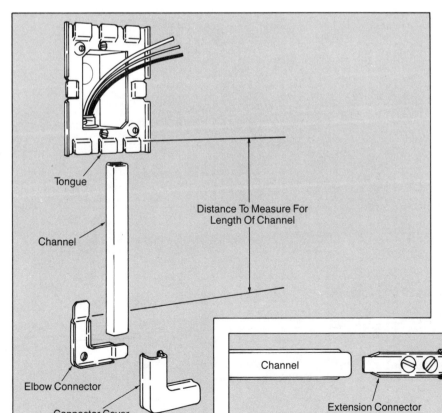

Tongue

Channel

Distance To Measure For Length Of Channel

Elbow Connector

Connector Cover

Channel

Extension Connector

Connector Cover

Installing a Raceway Channel.

Mark the wall, floor, or ceiling surface for your installation with a long straightedge or plumb-and-chalkline. First, install all corner and elbow pieces with the proper fasteners; however, attach them lightly. With a hacksaw (24 teeth per inch), cut each channel of the raceway precisely to the length between the base of the tongue of one unit to the base of the tongue of the next unit, as shown here. Be sure that pieces being cut are tightly secured or clamped and also be sure to ream any rough edges with a half-round file before assembly. If you make an error or come up short, you can connect two straight pieces with an extension connector to make the desired length. Install sections by pushing both ends onto their connecting tongues; it might be necessary to remove screws in order to insert some of the sections. Fish wires through the channels by first running a fish tape from one end to the next, then attaching the wires and pulling them through. Allow 8 inches at each end for wire connections. Do wiring at the outlets, then install covers.

Installing Outdoor Wiring

A well-lit home is warm in its greeting to nighttime guests—nothing highlights the beauty of your home more. But there are also practical reasons for adding outdoor wiring. If you use an electric lawn mower, power tools, or bug lights, an outdoor receptacle is an asset. Floodlights and electric-eye switches that activate lights at dusk make your home more secure from prowlers—plus they reveal icy or dangerous walkways.

This chapter will tell you how to install fixtures to remedy such problems, whether you install them on the outside of your house or in your yard. Although outdoor equipment differs from indoor, and is shown in detail, the wiring process is *not* shown because it is the same as for indoor installations (Chapters 4 and 5). An exception is the garden lights shown in Chapter 12; these require low-voltage wiring.

Begin by reading this entire chapter to gain an understanding of what is involved in each step. Then formulate an outdoor wiring plan. After checking into local code restrictions and choosing materials, start your project. The four main steps, in this order, are: (1) installing the source wiring, (2) digging the trench and laying cable or conduit, (3) installing boxes or fixtures along the wiring run, and (4) connecting the wiring. As always, safety procedures must be followed stringently when performing this final step.

Installing outdoor wiring isn't terribly tricky and doesn't require special skills; in fact it requires hard labor more than anything—in the digging for underground burial of the circuit. It might be wise to compare costs of hiring an electrician, doing it all yourself, hiring help to do the digging, and renting a trench digger. That way, you'll realize more clearly what your savings are.

Code Specifications

Before planning an outdoor project, first contact your local electrical inspector to find out the Code restrictions for your area. You might have to get your installation inspected when it is finished and you might need to notify your utility company that you're going to dig in your yard. You will most definitely have to install a Ground Fault Circuit Interrupter (GFCI) for any outdoor circuitry, and the Code will also have requirements about what kind of cable or conduit to use.

Guidelines for Materials and Methods. Almost all codes agree on the use of metal conduit in above-the-ground outdoor installations but they may disagree on whether rigid metal or EMT conduit (page 10) can be used. (EMT is thinner and less expensive, so you should use it where permitted to save costs.)

For underground burial, UF cable is preferred because it is flexible and much easier to work with—and most codes will permit it. If you do use it for burial, however, you must use conduit where the cable enters and exits the earth. Some codes require rigid nonmetallic (page 10) or rigid metal conduit as the buried material, depending on where the wiring run is located. If metal conduit is used throughout a run, then individual TW wires (page 9) can be used as conductors.

All methods have their advantages and disadvantages. Another factor to be considered is the depth of burial. UF cable must be buried at least 24 inches below grade, possibly more. Though trenches can be narrow, at this depth a large project, or even a small one, can be a back-breaker if tackled with a garden shovel. Renting a trenching machine could prove to be a wise alternative.

Rigid metal conduit is expensive and difficult to work with because it must be bent and connected at precise points in a wiring run—but its use will save you some digging. Most codes require that it be buried only 6 inches below the surface, but be sure to check yours for deviation.

Working with Rigid Metal Conduit

Though rigid metal conduit is hard to work with, there are some measures you can take to make an installation easier. First, in choosing materials, select aluminum conduit if your code permits you to do so. Some forms of surface protection may be necessary where this is buried in the earth. Always check the manufacturer's instructions. It is lighter and easier to bend, but provides the same protection as the galvanized steel type. Second, purchase or rent a *conduit bender*, a specialized tool that helps you to shape conduit.

In addition, make sure that you have these 'workshop' tools available: a round file to smooth rough edges, a C clamp or vise to anchor conduit while cutting, a hacksaw, a 12-inch-long adjustable wrench for fitting nuts onto threadless connectors, and two 10-inch pipe wrenches for screwing components together.

Work with conduit section by section, cutting and bending pieces as you need them. Assemble sections outside the trench you have dug. When substantial parts of the run are assembled, lower them into the trench and connect them to pieces already in place. Keep the

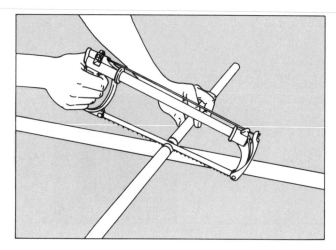

Cutting Rigid Metal Conduit. Hold the rigid metal conduit firmly in place and cut it with a fine-tooth hacksaw. First, wrap the pipe with tape to help prevent the saw from slipping. When finished, be sure to smooth the edges with a file.

runs as straight as possible. If you have many curves, add a box or components with access plates so that you won't meet resistance when fishing the wires. These must be marked permanently with a short stake or a fixture. Conduit run along a wall must be anchored with straps (page 10) every 10 feet or, when it leads to a box, within 3 feet of the box.

A Safety Essential— The GFCI

There is one thing that all local codes agree on and the *NEC* is adamant about: **Ground Fault Circuit Interrupters must be installed for all outdoor receptacles installed at grade level. They are your insurance against shock so do not neglect to make these installations.**

GFCIs (page 4) operate like the circuit breakers in your service panel except that these super-sensitive devices protect against extremely small amounts of electrical leakage. Normally a person's body will resist such electrical charges, except for a painful shock, but add to the scene a damp lawn or a sweaty hand and the results could be fatal.

Life-saving Ground Fault Circuit Interrupters are required for outdoor receptacles and recommended in all new outdoor installations, as well as in any location that is potentially wet. You may install one at the service entrance for the entire outdoor circuit or you may use a combination receptacle and GFCI fixture (shown right, center) at the first outlet on an outdoor circuit. Installed in this manner, it will protect all of the remaining outlets on the circuit.

BENDING CONDUIT

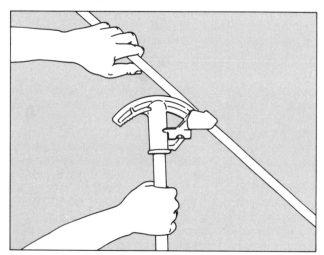

1 Place the conduit inside the bender at the hooked end, as shown.

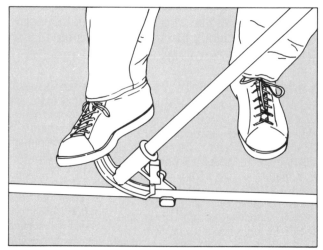

2 On a hard surface, put your foot on the footstep of the bender and pull towards you with the handle. The conduit will gradually bend to the desired angle.

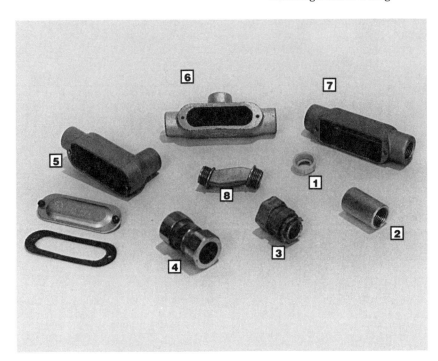

Rigid Metal Conduit Components.
These are the most commonly used components for typical outdoor wiring projects:

1 Plastic **insulated bushings** protect wires from damage.

2 **Threaded couplings** join threaded sections of rigid metal conduit.

3 **Threadless connectors** are used to connect outlet boxes, threaded couplings, and other components plus long or difficult-to-turn sections of conduit.

4 **Threadless couplings** join cut sections of rigid metal conduit.

5 **LB connectors** are for routing cable from an outside wall to the ground.

6 The **T body** is used for routing wires in two directions.

7 **C bodies** are for extending a run. Like LB connectors and T bodies, they have removable access plates to make wire fishing easier.

8 **Offsets** provide a way to jog past obstacles or redirect runs.

Because outdoor receptacle installations are in high-risk areas, GFCIs (Ground Fault Circuit Interrupters, page 4) are required. When installed at the first outlet on a branch circuit, they protect all of the succeeding outlets on a run. This fixture, a combination receptacle/GFCI has a leakproof gasket and cover and two push buttons. One (R) is a reset button to be used after the device has been tripped; the other (T) is for testing to make sure that it is operating properly.

INSTALLING A GFCI

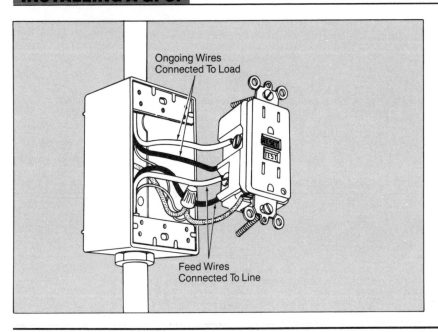

Ongoing Wires Connected To Load

Feed Wires Connected To Line

With the circuit power turned off, first test to determine which is the feed cable coming from the service panel. Connect the feed wires, black to black and white to white, to the terminals marked LINE. Connect the remaining cable wires, black to black and white to white, to the terminals marked LOAD. Grounding wires should all be spliced together and connected by a pigtail to the screw at the back of the box. If there is no ongoing cable, simply connect the LINE wires as described. If the GFCI has lead wires, connect the wires in the same basic way using wire nuts to make the connections.

Planning for Outdoor Circuits

First make a 'wish list' of the outdoor lighting fixtures and receptacles you want. Be careful not to overlight a patio or garden space and be sure to explore the possibility of using low-voltage lighting (page 111) for decorative purposes. Floodlights, entryway security lights, and lamppost lights, however, will all be run on general purpose circuits, as will receptacles.

Next, make a sketch of your home and trace imaginary wires from fixture to fixture. Remember that a lot of work and expense is involved, so make your runs as short and straight as possible. Avoid drain fields and any rocky areas where digging will be extra difficult. Also, avoid obstructions like sidewalks as much as possible and any spaces where other wiring might be, such as telephone cables.

In conjunction with making your sketch you will have to determine how many branch circuits are being created or tapped into and which existing power source(s) to use. If you are making a fairly elaborate, relatively high-voltage circuit, you should run a separate line from the main service entrance panel. If you are doing the same in a remote area, or if you also need more power in a detached garage or workshop, a practical step would be to install a subpanel. These are available in indoor and outdoor models.

Getting Power to the Outside. Generally, you will be able to tap into existing house wiring without overloading a circuit (page 51). You might be able to tap into an existing outdoor fixture such as a floodlight at a doorway; or you might find it convenient to tap into an attic circuit at an eave. Both of these, however, will entail running conduit alongside your house which might be unsuitable as an unattractive route.

Depending on whether or not you have a basement, you can direct wiring through a first-floor wall using back-to-back boxes, through a masonry foundation, or through a basement. At the exit from the house you can immediately direct the conduit toward the ground by using an LB fitting or you can install a GFCI/receptacle combination or a junction box. Many methods of getting wire outside are shown on the following pages.

Before selecting an outlet as a power source for your run, check to see whether it is suitable (page 56) and also whether it is part of a switch-controlled circuit.

Going Underground—Channeling Wires through the Earth. Once you have found and made your wiring exit, you can begin digging the trench for the circuit run (page 103). Whatever you do, don't dig the trench first because you may not be able to exit through a structure at the exact location intended.

Making the trench simply involves hard labor. If you use plastic sheeting and replace sod when finished, you shouldn't have too big of a mess in the end. As mentioned earlier, UF cable makes the channeling step less tricky. But even if you use conduit and conductors, you can correct errors and solve problems with special fittings. Still, if you are a beginner, you should work slowly and methodically.

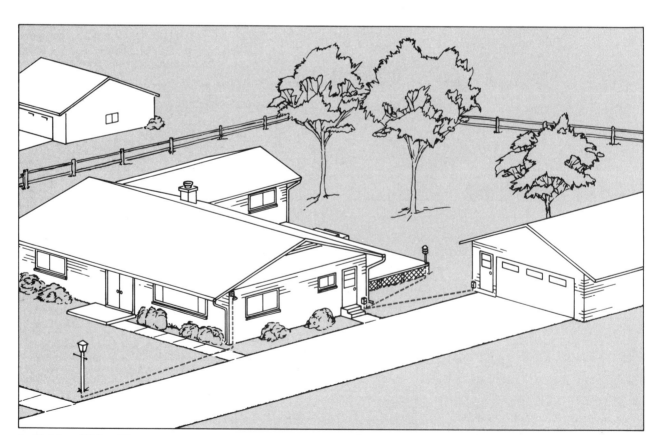

An Outdoor Wiring Plan. Here is a basic plan for outdoor wiring. It includes two circuits, one in the front of the house, one in the back. The front lamppost receives its power from an existing floodlight circuit. The backyard light and receptacle circuits branch out from a combination GFCI and junction box installed on the side of the house. Cable is run as straight as possible and crosses all obstructions such as sidewalks in the shortest possible distance.

INSTALLING A BACK-TO-BACK BOX INTO SIDING

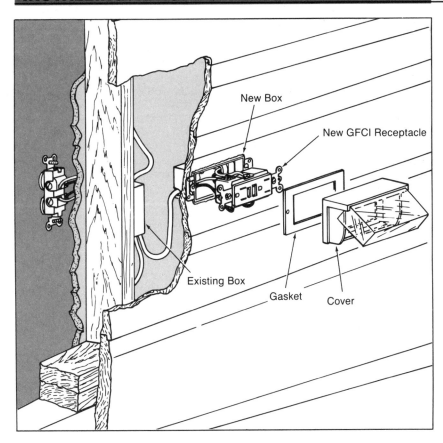

New Box

New GFCI Receptacle

Existing Box

Gasket   Cover

This installation is very similar to the indoor back-to-back installation on page 62 except that it involves cutting into wood, aluminum, or vinyl siding. Protect your eyes by wearing safety glasses or goggles. With the power to the circuit shut off at the service panel, remove the wall plate and receptacle from the box, leaving the wires intact. Punch out the knockout in the back of the box. Determine the new box position (page 102, step 1), measure the opening for the new outdoor box and drill ⅜-inch holes at each corner as starting points for the cavity. Cut through the exterior siding using a sabre saw with a metal cutting blade. Insert the cable through the knockout and push it in the direction of the new box. Have a helper pull it through. Install the new box, clamp cables in both boxes and connect all wiring. Patch the exterior wall with caulking compound.

INSTALLING A BOX INTO A MASONRY WALL

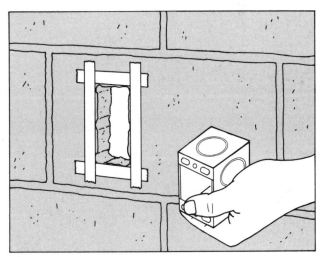

1 Position the hole wherever the block is hollow—for older concrete block, in the center; for new block, to one side. Never use the top row of blocks of a foundation— these are solid. Make an outline of the box with tape. Protect your eyes by wearing safety glasses or goggles. With a grounded or double-insulated electric drill fitted with a ½-inch masonry bit, drill holes and then break out any material left between the holes. With a masonry chisel or a special four-edged chisel called a *star drill*, chip away at the opening until the box will fit in it.

2 Make sure that you are using a special *concrete box* like the one shown here. Place the conduit and the box in position; the box should extend only about ¹⁄₁₆-inch from the wall (the space needed for the cover-plate gasket). Mortar it using a putty knife.

RUNNING CABLE OUTSIDE—FROM A BASEMENT

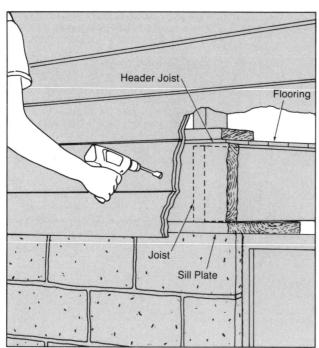

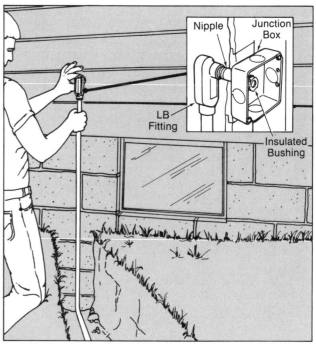

1 Determine the position of the cable exit. Use a reference point such as a corner or water pipe exit. If the wall is concrete block, do not use the top row of blocks which are solid; you must use at least the second row down. In a house with siding, the hole should be at least 3 inches from a joist, from the sill plate, and the floor. Hold an LB fitting to the wall where you intend to make the exit hole. The fitting should not overlap any joints in the siding. Make a ¼-inch test hole and check your measurements with a stiff piece of wire. With eye protection on, bore through the header joist with a ⅞-inch spade bit.

If the wall is masonry, use either a power drill with masonry bit or a ⅞-inch star drill to make the hole. Hammer the star drill after every ⅛-turn to make a neat, round hole.

2 Remove a rear knockout from a junction box and install it indoors over the hole you have drilled. If the interior wall is masonry, use plastic or lead screw anchors to fasten it. Measure for a nipple long enough to connect the box and LB fitting. Screw the nipple to the fitting and insert it into the hole. By now the trench should be dug for your underground cable run. Prepare conduit by bending the end which will lay in the trench. Connect conduit to the LB fitting and use a strap to attach it to the side of the house. Caulk around the nipple. Inside, secure the nipple to the junction box with a connector and a plastic bushing to protect the cable or wires from the metal edge of the conduit.

RUNNING CABLE OUTSIDE—FROM AN ATTIC

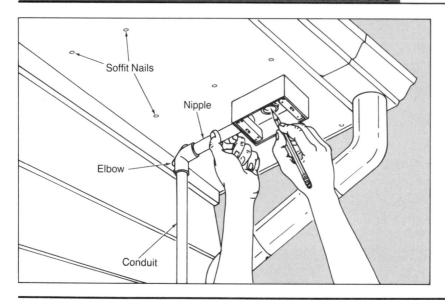

Turn off power to the circuit you are tapping. For this project you will need an outdoor box, a nipple, a corner elbow, and a section of conduit. Fasten all of them together and hold the assembly against the soffit between two rows of nails making sure that the conduit is snug against the house siding. Using the box as a template, mark the soffit for the cable hole and the screw holes for mounting. With eye protection on, drill holes using a ³⁄₃₂-inch bit for the ¾-inch, No. 8 mounting screws and a 1⅛-inch spade bit for the cable hole. Run cable from the power source indoors out through the hole in the soffit. Use a two-part connector (page 21) to connect it to the box. Mount the box, strap the nipple to the soffit and the conduit to the siding.

RUNNING CABLE UNDERGROUND

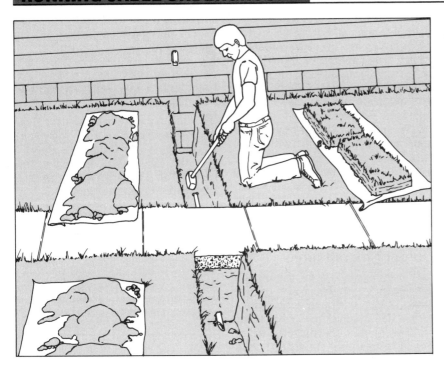

Once you have extended the cable out of doors, you can make your underground installation. Beginning at the point where the cable exits the house, make a line to your destination point with stakes and string. Beneath the string, mark the ground with chalk or flour. Dig a trench on this line—24 inches deep if using UF cable—no wider than a shovel, piling 3-inch-thick sod on one side of the trench and loose soil on the other. It is best to use plastic sheeting to hold the dirt, as your yard will look cleaner when finished. If you come to a sidewalk, put on eye protection and then break away as much of the bottom concrete as possible. Drive a pointed ½-inch steel pipe through what is left. The pipe should measure the width of the sidewalk plus a few inches more for extending on both sides of it. Use a sledgehammer to drive the pipe. When the ditch is ready, lay out all the components of your run: conduit, fixtures, and fittings. Make sure that everything matches and then run cable through straight conduit. Next run it through any bent conduit, make connections and do wiring.

WEATHERPROOF FIXTURES

The heavy-duty metal box (upper left) is made specifically to resist weather and moisture although it is not completely waterproof. Shown are three types of covers for installing outdoor lights, receptacles and switches. If you have to install a box in an area subject to flooding or spraying by sprinklers, you must use even heavier, more expensive watertight fixtures.

Outdoor Light Fixture. Shown here is a basic outdoor box, gasket and cover; other sizes and shapes are available. The fixture is connected to the cover plate with a star nut and the socket contains a gasket to seal the gap inside. All outdoor lights require especially made weatherproof bulbs that will not shatter with a sudden temperature change.

Outdoor Receptacle Box. Can be used with regular receptacles. Contains a cover-plate gasket inside and a cover plate with two doors. The doors have gaskets and are loaded with springs to close when not in use. Openings for conduit are covered by threaded plugs.

Outdoor Switch Box. A preassembled switch and cover plate fit over an outdoor box to complete this fixture. The small handle on the outside controls the toggle switch inside the box. Also has threaded plugs that close unless used for conduit.

Special Fixtures for Outdoor Use

Once you have completed the steps of exiting cable or conductors from the power source and burying the cable or conduit, you can hook up your new fixtures. Outdoor electrical fixtures are wired in the same manner as regular indoor fixtures, but in their construction they differ considerably. They are made of heavier metal to resist wear and rust and all openings are protected against moisture by gaskets. Instead of knockouts, they contain threaded holes into which the ends of conduit or metal plugs are screwed.

There are several types of outdoor fixtures and the kind shown on page 103 will give you standard, minimum protection from 'average' weather. But depending on where you are going to mount the fixture and where you live, you might need to use better-sealed watertight fixtures.

Shown are the three basic fixtures most commonly used: a receptacle, switch, and light. Also, at right is a specialized fixture, an electric-eye switch, which gives you convenience and security benefits. This gadget, which comes in several styles, measures the amount of light outdoors and, when the level is low, automatically switches on your safety and security lights.

When installed at the beginning of a run, an electric-eye switch will control other lights on the circuit. Also, the device itself can be controlled by a switch, for times when you don't want it to work. Many models are available. Though different, they are all wired basically the same.

Outdoor Installations

Here are instructions for mounting a lamppost and a box (both end-of-run and middle-of-run) along your buried wiring path. After making these installations or other outdoor installations, follow these steps to finish your project:

1 Once you have pulled wire through to begin the wiring, make sure that the current is dead. To protect your eyes, put on safety glasses or goggles.

2 Make the wire connections.

3 Restore power and use a voltage tester to check that the circuit is operating properly.

4 Fill in the trench you have dug.

For protection of a UF cable run, you can place redwood or treated boards over the cable before backfilling the trench. This is an extra measure against damage due to future digging.

WARNING

Always turn off the current before making any wire connections by shutting down the circuit at the service entrance. *Always important, this step is even more critical in outdoor wiring where damp ground makes your body more vulnerable to electric shock.*

INSTALLING A LAMPPOST

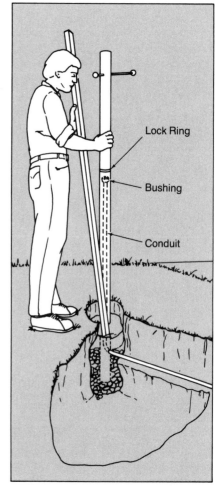

Lock Ring

Bushing

Conduit

1 Lampposts need to be buried at least 2 feet deep. If you are using UF cable you can use the access hole provided on the post for cable entry but if you are using conduit you must first cut a channel for slipping the post over the conduit. If the post is at the end of a wiring run, you only need to cut one channel; if the wiring continues, you need to cut two. Mark for the channel, measuring two 18-inch lines spaced ⅞ inch apart. Lay the post down, secure it and begin cutting with a special small hacksaw. (A regular hacksaw will not work because the frame will get in the way.) When finished, bend back the piece and saw it off. File down all sharp edges. If necessary, repeat the process on the other side of the post.

2 Dig a hole 2 feet deep and approximately 8 inches in diameter. If you are using conduit, measure, cut, and bend a piece that will run from the existing conduit into the channel and up through the lamppost. If your lamppole is adjustable, cut the conduit to go to the height of the lock ring; if it's not, cut it to run the entire length of the post. Add a nonmetallic bushing on the end to protect wiring. Slide the lamppost over the conduit (if using UF cable, simply mount it and feed the cable through). With a 2 x 4, tamp layers of dirt and large stones into the hole. Periodically check the post for vertical alignment and fill the hole only to the bottom of the trench. The hole can be filled later after wires are fished and the trench is filled.

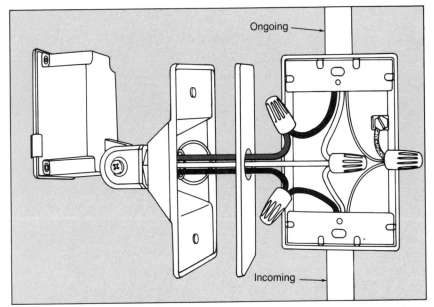

These electric eye switches operate on a tiny current to turn lights on when it gets dark outside. Two models are shown. One comes with an outdoor cover plate and nut so that only the sensor is exposed to the weather. The other kind has a raintight body. It is simply screwed into an outdoor lamp-holder cover and can be rotated to point up or down as needed for light sensitivity. Both switches have lead wires and are easily hooked up.

Wiring an Electric Eye Switch. Install this switch at the beginning of a circuit run of lights to control any number of lights up to the capacity of the switch. With the circuit turned off, and with eye protection on, connect the three wires of an electric eye switch as follows: With a wire nut connect the white wire to the white wires in the box. Connect the black switch wire with a wire nut to the black wire from the incoming feed cable. Connect the red switch wire to the other (ongoing) black wire. Attach the grounding wires together with a wire nut and run a green pigtail to the back of the box.

INSTALLING OUTDOOR FIXTURES

Installing an End-of-the-Run Box.
After you have dug your trench and deter-mined the position for this light or receptacle, measure for a piece of conduit that will rise out of a concrete building block. The block should be approximately 8 inches below the ground surface and the fixture should be tall enough so that it won't be tripped over or sub-jected to extreme weather. If you are using UF cable for your run, attach the cable to conduit with a bushing. Pull wiring from the conduit and mount the box either by screw-ing it on or using a threadless connector. Place the block over it and straighten the conduit, using a level, until it is vertical. Fill the cavity of the building block with mortar. Wire the outdoor box for a light or receptacle.

Installing a Middle-of-the-Run Box.
This installation is very similar to the previous one except that you will need two bent pieces of conduit and either an outdoor junction box (as shown here) or a receptacle box with two bottom knockouts. Whatever box you choose, make sure that it is large enough to hold the wiring (page 74). Snake wiring through both pieces of conduit and into the box. Attach the box by screwing it on or using threadless connectors. Position the conduit until it is ver-tical and pour the concrete. If using a junction box, connect wires with wire nuts—black to black, white to white, and ground to ground.

Low-Voltage Wiring

There are many low-voltage products available to the do-it-yourselfer for home modernization. Integrated circuits, remote control, infra-red systems—all have contributed to the new age of electronic magic. The wave of the future is even more amazing; someday, through the use of computers and computerized appliances, many of our everyday chores might be performed automatically.

In this chapter you'll learn some basics of low-voltage wiring through three popular subjects. You'll learn to diagnose a silent doorbell and how to install new door chimes at two locations.

There is a stereo speaker installation for the room that you've always wanted to fill with music; and with a separate volume control you can keep sound at different levels in each room. These two projects require concealed wiring, covered in Chapter 7.

Low-voltage yard lights are covered. They give your yard an enchanting glow and involve minimal labor because wires can be placed just beneath the ground surface. Try out or at least acquaint yourself with these projects—to see how simple and exciting the wiring can be. Next you'll want to learn soldering or put in your own intercom system. You might even become a 'tinketeer'!

Stepping Down the Power

Low voltage is defined as being from 20 to 30 volts or less—a current flow-pressure significantly lower than the standard 120 volts with which most homes are wired. Because it is so low, working with low voltage wiring is not dangerous and there are few Code restrictions; the worst low voltage shock would barely tingle your hand. There are a few exceptions: **If you are wearing a pacemaker do not attempt any wiring.** Also, installing a *transformer,* which many systems require, is dangerous. This device steps down your house power to the low voltage needed for the device you are installing. The current must be shut off when you install a transformer.

There are several types of transformers. Often they will simply be built into a device; at other times they are separate units that are plugged into a receptacle. The ones shown here for door chimes and outdoor lights are wired into 120-volt circuits at junction boxes. Regardless, they all have this stipulation: Transformers cannot take excessive heat. Do not install them in hot attics or tightly confined spaces, and keep them away from insulation or high-wattage lights. Also, never install a transformer inside of a service panel board.

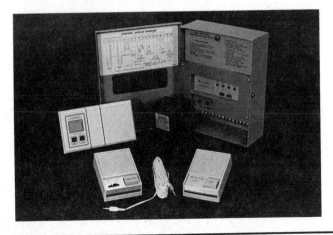

These are just a few of the many low-voltage products available in home electronics supply stores. Shown are, counter-clockwise: a security alarm system, a two-station intercom, and a programmable thermostat. All come with easy how-to-assemble instructions and diagrams.

INSTALLING LOW-VOLTAGE DOOR CHIMES

Doorbell or chime installations require 18- to 20-gauge 'bell' wires (page 9). Grounding wires are unnecessary and since there are no 'hot' or 'neutral' wires, any colors can be used. For simplicity, however, purchase three different colors of wire when installing a system like this one, with two chimes. Buy wire for the length of the run plus an extra 15 percent.

Transformers are installed most easily in basements and they must be connected to a regular outlet box to reduce the current from 120 volts to the low voltage required for the system (typically 16 to 20 volts). The circuit power must be shut off when making this installation. First, extend a circuit for this purpose (page 56); it cannot be operated by a switch. Chimes are often installed in entries or hallways. Bell wires are permitted by Code to be exposed but most people prefer to conceal them behind walls. Use the methods on pages 57-63 for fishing wires in finished spaces. Study the circuit pathways in this illustration and then follow these steps for the installation:

1 Drill holes for the push buttons.

2 Fish wires from the transformer to each push button.

3 Run wires from push buttons to chimes.

4 Run a wire from the chimes to the transformer.

5 Connect the chime wires.

6 Turn off the power at the service panel to the branch circuit that the transformer is connected to.

7 Connect the transformer to the house circuit.

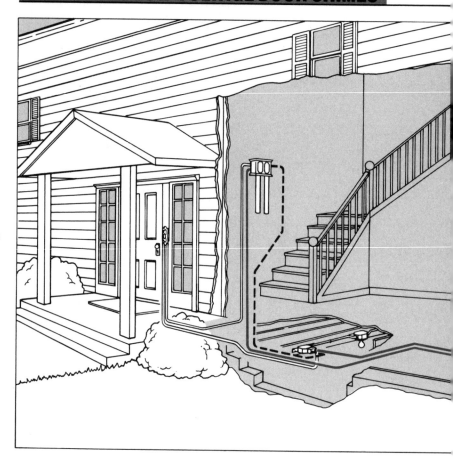

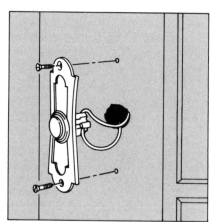

Mounting the Push Buttons. In the hollow space of the outer wall of the house, about 4½ inches from the outer edge of the door, bore a ⅝-inch hole. Position the button as for wiring; mark and drill the pilot holes for mounting screws. After wires are fished through—one from the chimes and one from the transformer—connect wires to the screw terminals and mount the button.

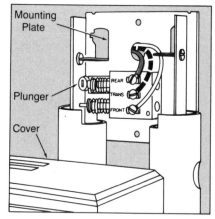

Installing the Chimes. With the mounting plate, locate and mark holes for wires and screws. Drill the holes and after fishing three wires through the wiring hole—one from the transformer and one from each push button—mount the chimes on the wall. Make wire connections to screw terminals. Connect the transformer wire to the trans terminal, the rear chime wire to the rear terminal, and the front chime wire to the front terminal. Make sure the wires are out of the way of the plungers.

TROUBLESHOOTING FOR DOOR CHIMES

1 Test the door chime button by removing it, scraping away any corrosion, and jumping the terminals with a piece of wire. If the chimes sound, the button is faulty and you need to replace it. If they don't sound, disconnect the wiring on the button, twist the wires together, and proceed with step 2. (The door chime will automatically sound, saving you steps running back to test it).

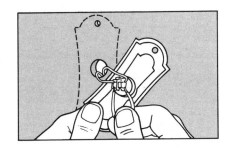

2 Check the transformer which is probably attached to a junction box in the basement or located elsewhere close to the chimes. If one of the wires is loose, reconnect it and the chimes will sound. If not, proceed with step 3.

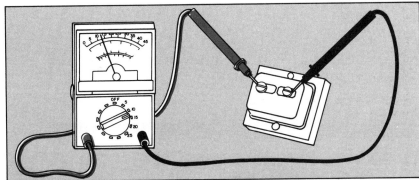

3 Test the transformer by removing the low voltage wires and probing the terminals with a special tester called a voltmeter. If you get a zero reading, you must replace the transformer.
If the transformer passes the test, proceed with step 4.

WARNING

Make sure that the branch circuit is shut down at the service entrance before removing the wires supplying power to the transformer.

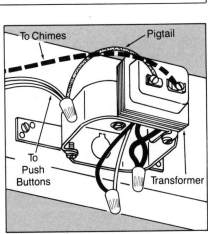
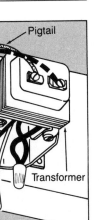

To Chimes   Pigtail

To Push Buttons   Transformer

Making Connections at the Transformer. With the branch circuit shut down, extend a cable from the house wiring and mount a junction box on a floor joist in the basement. Remove one of the knockouts, insert the transformer wires into the box, and attach the transformer to it using the 'quick clamp' that is attached to the transformer. With wire nuts connect one of the transformer wires to the black house wire; connect the other transformer wire to the white house wire. Attach the ground wire to the box. Connect the wire from the chimes to one of the terminals on the transformer. To the other terminal connect a pigtail which has been spliced with both of the wires from the push buttons. The pigtail can be longer than shown here, branching off from a convenient spot where push button wires meet.

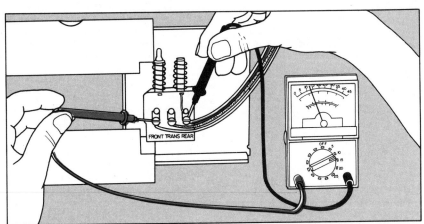

4 Examine the chime unit for loose connections. To test a two-button system, with wires intact, place voltmeter probes to the front and trans terminals. If it registers voltage, the chimes are broken. To test the rear chimes, probe the rear and trans terminals. To confirm that the fault is not in the chimes, disconnect the chimes' leads and test for currents with a voltmeter. If all components check out, then you know that the problem is a broken wire along the wiring run. You will have to either trace and locate the problem through the walls of your home or install a new system.

INSTALLING AN EXTRA SET OF STEREO SPEAKERS

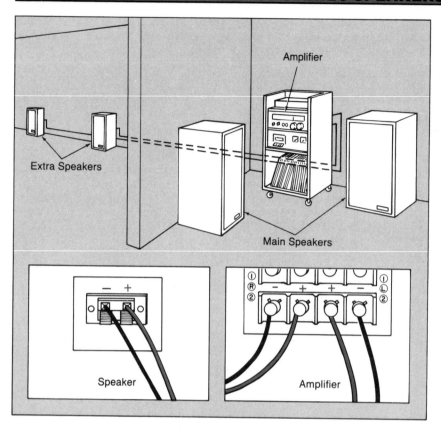

Check your amplifier for an extra set of terminals for a second set of speakers. Run color-coded, two-conductor cables—No. 18 for a 60-foot run, No. 16 for a longer run—from the amplifier to each speaker. Strip away ½-inch of insulation and insert conductors into the spring-loaded clamps of the speakers, noting which color conductors are matched with the positive and negative terminals. At the amplifier, connect the positive conductor from the right speaker to the amplifier's right, positive terminal. It will be marked either by a plus sign or a red dot. Connect the other conductor to the right, negative terminal. Repeat the process to connect conductors for the left speaker. Variations exist in speaker and amplifier terminals. Some brands have screw terminals which accept 'tinned' or soldered wires. Though not necessary, the above installation is more secure if wires are soldered. Others accept spade lugs or phone plugs.

INSTALLING AN EXTRA VOLUME CONTROL

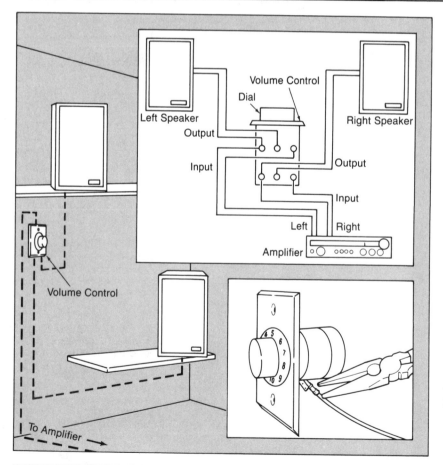

1 For this project use standard speaker cable. Usually clear colored, it will contain two differently-colored wires. The wires might be gold and silver or they might be gold and gold and white striped. In an area at least 12 inches away from any existing wiring, cut a rectangular hole large enough to accept a standard wall box. From this wall box run four cables—two cables to the amplifier and one cable to each speaker. Mark the cables in the box so that you'll know what they all lead to. Make connections at the amplifier and speakers using the terminals provided and the diagram shown here. (Products might vary so be sure to check the manufacturer's directions as well.)

2 With the wiring pulled through the box, make the connections at the back of the volume control. There will be six terminals and they will be color-coded to aid you in making the correct connections. Shown here are easy-to-use clamp-type terminals. Simply twist the bare wire strands of each wire, insert it into the clamp and crimp it tightly together with needlenose pliers. Attach the faceplate to the box.

INSTALLING LOW-VOLTAGE OUTDOOR LIGHTS

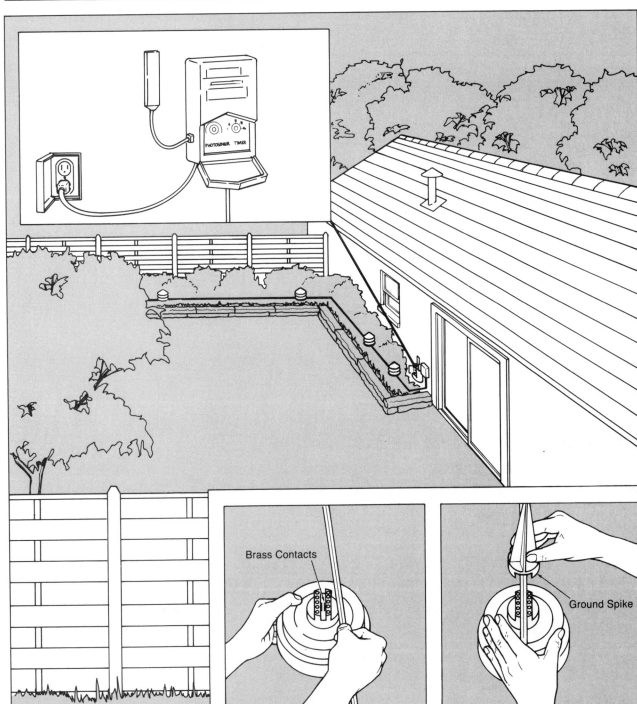

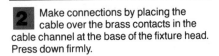

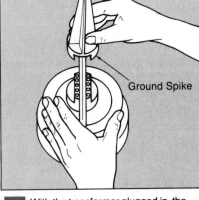

Brass Contacts

Ground Spike

1 These low-voltage 'mushroom-style' accent lights are easy to install because the cables don't have to be buried as deeply as regular outdoor cables. They are purchased in kits complete with instructions and sized by the number of lights. The kit shown contains a plug-in transformer, a timer, a photoelectric cell sensor, lighting fixtures, and all the wiring between fixtures. Mount the transformer and censor near a standard outlet. Then mount the fixtures; dig shallow trenches at least 6 inches deep with an edging tool. Lay the wiring leaving slack throughout the run and at each fixture. Make wire connections either with pin connectors or wire leads.

2 Make connections by placing the cable over the brass contacts in the cable channel at the base of the fixture head. Press down firmly.

3 With the transformer plugged in, the timer set to ON, and the sensor shielded from light, secure the connections by sliding the ground spike into position, as shown. The lamp should now light. Make a hole about 8 inches deep into the ground at the desired location (use a sharp gardening instrument for this step). Insert the spike into the hole; do not hammer on the lamphead assembly.

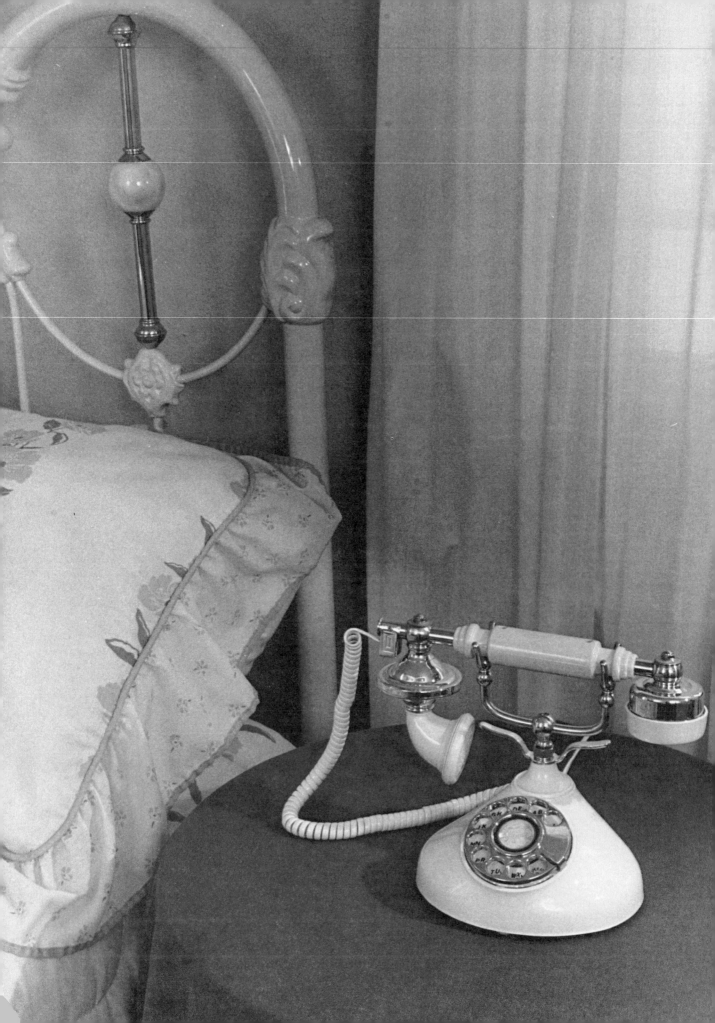

13

Doing Your Own Telephone Wiring

It's a truly marvelous era for communications. We have wonderful devices that take messages, accept computerized information, and notify us when another party is calling. We're able to make overseas calls with the push of one button, and there are cordless telephones that travel along with us.

At one time our entire telephone systems were under the domain of the telephone company. Any alterations or additions were conducted by them. Consumers had the choice of paying high prices for installation or having 'underphoned' homes. Then, a breakthrough. With the FCC's 1977 ruling, individuals were given the right to buy and install their own telephones.

The cost savings are substantial. By purchasing your own equipment, you can typically begin receiving an investment return in two years. By doing your own installations, you'll save even more. The prevalence of home electronics stores makes this work comparable to other household projects. Many components are available for updating systems so that easy modular hookups can be made.

This chapter will tell you how to assess what you now have, and it will tell you what part of your system you can work on. Telephones are considered low-voltage installations but there are some safety procedures that you must follow. You can use the techniques in Chapters 6 and 7 for mapping runs and hiding wires in the walls of your home—or you can make easier baseboard runs. Best of all, the wiring is extremely simple—it's all color-coded. Also, there are no specific tools that need to be purchased for these projects. So read the chapter thoroughly and plan for a new phone and determine your needs—you're on your way to saving money!

Telephone Basics

Telephone wiring can be amazingly simple but there are a few things that you should first understand about how the system and how the phone itself works. Whether service arrives at your home via underground or overhead cables, it will halt at a box outside your house called the *station protector*. This grounding and safety device should not be tampered with when doing installations. However, locating it on the building can help you to determine where the wiring enters your home.

Once inside the home, a two-conductor telephone cable is run to an initial termination called a *terminal block*. There are several styles of terminal blocks although the most prevalent is the *connecting block*, or, as it is known in the trade, the *42A block*. This is the point at which you, the resident, have control over your phone system. The only exceptions are systems with a party line or a pay phone (then only the phone company can do the interior phone work).

Depending on where you live and how advanced your supplier is, you will have an old or modern system to comply with and outdated or modern equipment in place within your home. Old-style *pulse-dialing*, typified by the rotary dials on old black telephones, sent pulses to the telephone company; these had to be stored and counted. Modern *tone-dialing*, such as that on push-button phones, sends much faster signals and has increased the phone company's time efficiency by a whopping 92%!

The equipment in your home also varies according to locale and date of installation. Phone 'outlets' throughout your home might be permanently wired or contain four-prong jacks. They might be connecting blocks or the new modular jacks that come in a variety of sizes and shapes. Again, the most common of the older models is the connecting block.

The main objective for the home do-it-yourselfer is to convert your old system to a new modular system. Once this is accomplished you can install new modular phones by simply plugging them in. (Even wall telephones are plugged in—along with the mounting process.)

Converting to a modular system isn't as complicated as you might think, because there are so many conversion components available and they are so easily connected to old systems. Shown here is a quick conversion from a connecting block to a modular jack, a project involving no actual wiring. Kits with complete instructions are available for other types of conversions.

Safety Precautions. As mentioned in the previous chapter on low-voltage wiring, do not attempt any wiring if you are wearing a pacemaker. Also, be sure to avoid working in wet areas. Most importantly, observe the warnings on these two pages which apply specifically to telephone wiring.

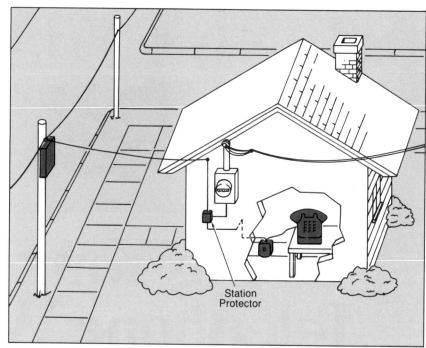

Telephone Wires Outside the Home. The telephone company sends out cables which terminate in boxes located on utility poles. The two-wire cables branch off and are strung to individual homes. Whether your service comes from overhead, as shown, or underground, the telephone cable culminates at a *station protector* before entering your home. This safety and grounding device protects your system from high voltage power surges that occur during thunderstorms. The station protector is the responsibility of the phone company; it should not be worked on except by trained personnel.

CONVERTING A CONNECTING BLOCK TO A MODULAR JACK

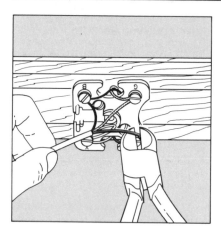

1 The very easiest way to change a connecting block over to a modular jack so that you can plug in new-style telephones is to purchase a conversion or 'instant' jack. If you are removing an old rotary-dial phone and installing a modern one, you should first contact your telephone company and ask to be connected to tone-dialing service. To remove a phone from the connecting block, first remove the handset so that the phone is off the hook. Next, remove and discard the cover of the connecting block. Inside, snip off the wires that lead to the old telephone. Do not snip any other wires or loosen any screws.

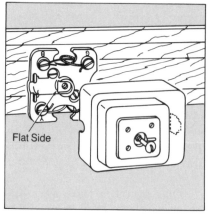

2 Line up the conversion jack with the block, as shown. There is a ring in the center of the block which has a flat side; this side should be opposite the new modular jack's opening (where the phone will be plugged in).

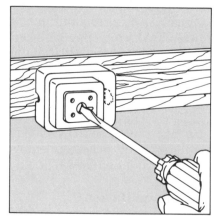

3 Place the new cover over the assembly and screw it down very tightly to make the 'automatic' connections. These connections are made by spring finger contacts inside the conversion jack.

WARNING

If the phone you are removing has a light in it, there is an additional transformer supplying the phone with AC voltage. Locate it and unplug it before snipping any wires.

The ringing of a telephone increases its electrical charge and makes it dangerous to work on. When removing phones and contacting the wiring, *take the handset off the hook of either the phone you are working on or another phone in the house.* Ignore the recorded message that tells you to hang up. This step notifies the central office that the phone is in use and prohibits calls from coming in.

Telephone Wire. There are two basic kinds of telephone wire: cable for wiring runs and telephone set wiring—'extension' wire used to connect modular phones. Telephone cable, usually in 22-24 AWG size, comes in rounded or flat, rectangular form and has a solid-color outer sheathing. It is handled like other wires—stripped and spliced. Extension wire is a silver-satin color and comes pre-measured in either plug/plug style or plug/spade style. Plugs and spades are provided for easy connections but, if necessary, the wire ends may be cut and spliced and connected like cable wires.

The interior of both kinds of wire consists of four color-coded conductors (sometimes there will be only two). Connections are extremely easy in home phone hookups because everything is color-coded. The primary conductors that cause your phone to work are called the *ring and tip conductors.* The ring conductor is always red and the tip conductor is always green. The other two conductors are yellow and black. These four separate wires are always connected to color-marked terminals, sometimes abbreviated: R, G, Y, and B.

The only deviations from this rule are in phones that have L1 and L2 terminals. In this case, connect the red conductor to L2 and the green to L1. Also, very infrequently a yellow conductor should be connected to the same terminal as the green conductor, but in this instance the phone manufacturer will give explicit instructions.

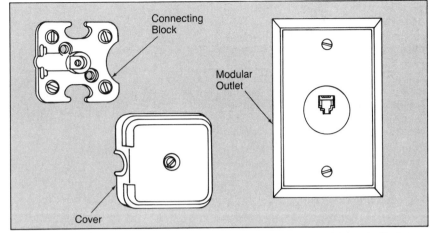

The Telephone Line Connection Inside the Home. After the telephone line enters your home, it terminates in a terminal block. This fixture is usually accessible, and though not all types are shown here, it will probably be one of these two: a connecting (42A) block or a modular outlet. If what you find in your home does not look like what's shown here, check your telephone directory which might have diagrams showing other types of terminals.

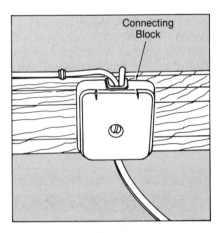

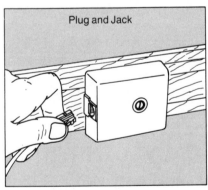

Telephone Set Connections Throughout Your Home. After the terminal block, the next element of telephone wiring is the telephone set which is the last connection that your telephone is 'plugged' into. These fall into two main categories: old-style and modern. The older style connectors include connecting blocks and four-prong jacks. There is also a permanently wired connecting system (not shown). The modern method features a modular system that uses plugs and jacks. This is the preferred connection since all new telephones are equipped with matching plugs.

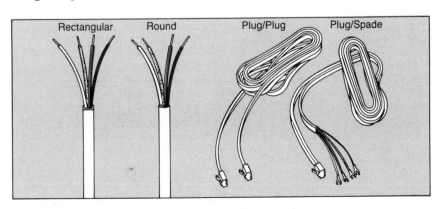

Types of Telephone Wiring. Shown here are cables used to make telephone wiring runs and telephone set wires (wires used between connections and phones). Flat rectangular and round cables, when spliced, reveal four conductors—red, green, yellow, and black. Telephone set wires, plug/plug and plug/spade types also contain the same number and color of conductors. The conducting wires are color-coded to match the marked screw terminals inside the telephone or modular jack.

Telephone Installations

As with regular wiring, the most difficult part of the job is often the planning, preparation, and carpentry work. First, decide how many and what kind of telephones you want to install and whether you want to keep existing phones in place or move them to more convenient areas.

As you can see in the circuit diagrams here, phones can be installed in various arrangements—with connections made either to the connecting block or to each other. A sample run is also shown with complete wiring and safety steps that should be followed for any new installation.

Make a plan of your home and review Chapters 6 and 7 for ideas on how to run cables through the walls of your home. A simpler method is surface wiring which is acceptable with low-voltage telephone wire; but you might prefer the neater look of concealed wires and faceplates that are set into walls. If you do run wiring along your floor and under carpeting, make sure to do it correctly, as shown here. A method used often by phone companies, especially practical for bedroom installations, is to run wire through closets.

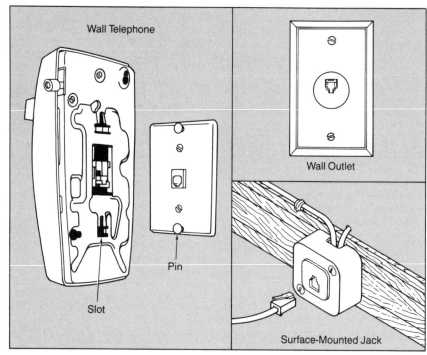

Types of Modular Connections. Besides the conversion jack shown on page 114, there are three other basic types of modular connections. The wall telephone connection is made by a modular jack mounted into the wall and the telephone plug which is inserted into it. The mounting plate also contains pins which catch slots on the phone backing and secure it in place. The wall outlet is built into the wall and contains jacks for plug-in telephones. The surface-mounted jack is the same but since it is mounted on the surface, it allows for exposed wires leading to other extension phones.

BASIC TYPES OF TELEPHONE INTERCONNECTIONS

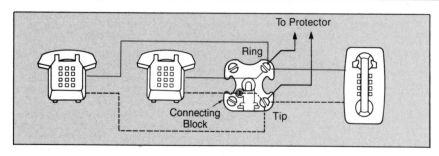

These three diagrams show a connecting block with ring (red) and tip (green) wires connecting to the telephones. The telephones are in different positions in relation to their power source.

Wire Pairs Starting at a Junction Point. In this configuration, all three phones are connected at the connecting block.

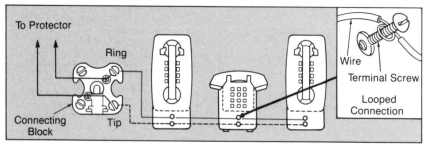

Connections Made from One Telephone to the Next. The first telephone on this circuit is connected to the connecting block. The second one is connected to the first telephone and the third connected to the second. This is sometimes called a 'looped' connection.

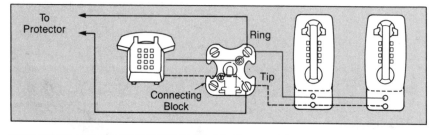

Combination Connections. This diagram shows a combination of the first two kinds of connections. The desk and first wall phone are both connected at the connecting block, while the second wall phone is connected to the first wall phone.

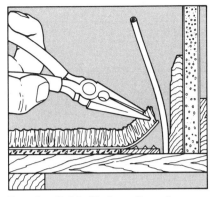

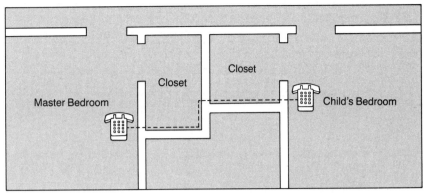

Running Cable Under a Carpet.
Telephone extension cords may be run
under carpets. Use needlenose pliers to pull
up the extreme edge of the carpet, without lift-
ing it off the inner edge of the tack strip. Push
the wire down into the cavity, being careful
not to place it on top of the tack strip. Press
the carpet back into place.

A Closet Run. Back-to-back closets, like
those shown here, provide easy routes for
telephone installations. Using them saves
you in the cost of a shorter extension cord
plus gives you a neater, finished look with
hidden cords.

EXTENDING FROM AN OUTLET FOR TWO TELEPHONES

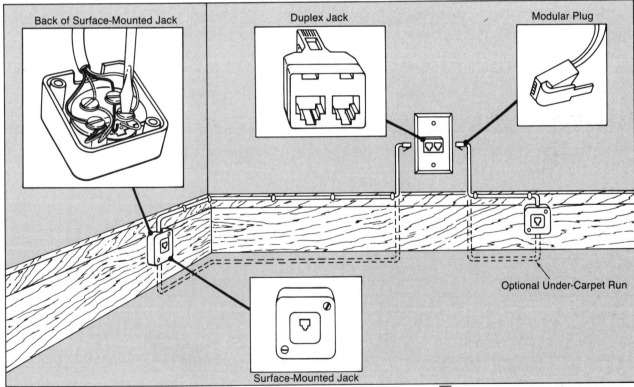

Back of Surface-Mounted Jack

Duplex Jack

Modular Plug

Optional Under-Carpet Run

Surface-Mounted Jack

Though this diagram is shown in one room
for brevity, use this method to place phones
in two separate rooms and extend the circuits to the lengths you prefer. Although the
baseboard method is pictured, you may also
run the cord under the carpet.

1 Beginning at the present modular outlet,
insert a duplex jack. Into it, plug the modular
end of one of the 'modular-to-spade-lug'
extension cords.

2 Drop the cord to the baseboard and
secure it with an insulated staple. Run the
cord along the baseboard to the first location
for your first new telephone.

3 Unplug the duplex jack.

4 Break out a slot in the surface-mounted
jack for the extension cord wire to exit at the
top of the baseboard. Connect the spade
lugs to the surface-mounted jack or, if the
cord is too long, cut it, strip the wires, and
connect bare wires. Connections are color-
coded—red to red, green to green, yellow to
yellow, and black to black.

5 Screw the surface-mounted jack to the
baseboard. Staple the cord to the baseboard
at periodic intervals.

6 Reinsert the duplex jack and plug the
modular end of the other extension cord into
it. Repeat steps 2 through 5.

7 Plug in the duplex jack and plug in the
new telephones.

APPENDIX 1: TROUBLESHOOTING HOME WIRING

TROUBLE	POSSIBLE CAUSE	SOLUTION
Fuse repeatedly blows or circuit trips often.	Faulty appliance or appliance cord.	**1** Put a 25-watt bulb in the fuse socket of the problem circuit. If it lights, there is a short circuit. One at a time, unplug appliances and turn off lights. When faulty appliance is unplugged or light is switched off—the bulb in socket will burn dimly. Check the appliance cord and replace if necessary. Repair or replace appliance. (If you have circuit breakers at the service entrance, use the same method and when the breaker trips you will have found the problem.)
	Short circuit or damaged wiring.	**2** If step 1 fails to show an appliance problem, you need to check all wiring on the circuit or call an electrician.
	Overloaded circuit.	**3** Calculate the load on the circuit (page 51) and then either move appliances (or possibly one large appliance) to another circuit. You might need to run a new circuit for the appliance.
Lamp won't work.	Bad light bulb.	**1** Screw in a new bulb that you have tested in another (good) socket.
	Circuit breaker tripped or fuse blown.	**2** Check the circuit at the service entrance. Replace fuse or reset the circuit breaker.
	Faulty receptacle (or switch).	**3** Check receptacle with a voltage tester; insert probes into slots of receptacle (page 14). Check switch by removing it from wall and testing (page 15). If bulb does not glow, receptacle (or switch) should be replaced and you should also check the wiring.
	Faulty home wiring.	**4** If testing in step 3 above failed, remove the wall cover plate and test circuitry by placing one probe of the voltage tester to a black wire, the other to a white wire or grounded metal box. If no light glows, the wiring is faulty and should be inspected along the circuit.
	Lamp is broken.	**5** If tests in steps 1 through 3 are successful, then the problem must be in the lamp. Test and repair using instructions on pages 41-43.
Switch fails to turn on light.	Bulb is blown.	**1** Replace bulb with a good bulb that lights in another socket.
	Fuse is blown or circuit breaker is tripped.	**2** Check at the service entrance; replace fuse or reset circuit breaker.
	Wiring at the light fixture is faulty.	**3** Test with a voltage tester (page 14); correct the wiring.
	Wiring at the switch is faulty.	**4** Use a voltage tester to check the wiring (page 15); correct if it is faulty.
	Switch is broken.	**5** Remove switch from the outlet box and test (page 15). Replace if necesssary.
Outlet faceplate is warm.	Amperage demand on circuit is too high.	**1** Check appliances or fixtures hooked up to outlet. If it is a lighting fixture, check to make sure that you are using correct wattage bulbs. Find the circuit and calculate the load; you might have too many high-powered appliances on the circuit. If so, move appliances or install a new circuit.
	Wiring is faulty at the outlet.	**2** Remove the faceplate and inspect the wiring. Look for bare wires touching each other and make sure that the wires are properly attached to terminals.
	Incorrect size wiring.	**3** Note the size of the conductors and the cable. Consider what appliances are being run on the circuit and replace wiring if necessary. Other remedies are to move appliances or install a new circuit.

APPENDIX 2: BUDGETING WIRING RUNS

A licensed contractor or electrician would use a set figure for each outlet installed and would simply multiply that figure times the amount of outlets. For example:

10, 120-volt outlets x $12.00 per outlet = $120.00

6, 240-volt outlets x $80.00 per outlet = $480.00

4, 120/240-volt outlets x $85.00 per outlet = $340.00

These are estimates only and will vary greatly according to locale; they are presented only to illustrate the method used. Electricians also have set figures for service entrance wiring, and they have average allowances for fixtures that would vary according to the owner's preference.

When budgeting for doing your own project, you can generally be more accurate because you will have measured the wiring run which, unless obstacles are encountered, will be completed as planned. Use the guide below to figure the estimated cost for your project. This guide is for a typical indoor project. When doing surface wiring, outdoor wiring, or low-voltage wiring, be sure to include such obvious needs as conduit and special fixtures.

Fill out a budgeting sheet, doing most of your research by phone to determine prices. Certain electrical supplies might be more expensive at a hardware store than at an electrical supply store. Determine where your best bargains are without sacrificing quality. Always purchase UL marked products. Try to purchase all of your supplies on one shopping trip to avoid interruptions while making an installation.

Tools

Tools for wiring	$_____
Tools for dismantling	$_____
Tools for repair work	$_____
SUBTOTAL	$_____

Fixtures

Ceiling lighting fixtures	$_____
Wall lighting fixtures	$_____
Switches	$_____
Special switches	$_____
Receptacles	$_____
Special receptacles	$_____
SUBTOTAL	$_____

Wiring

Cable—amount determined by measuring distance between each run (outlet to outlet), adding an extra 12 inches for each cable connection, plus adding an extra 10-20 percent to total for errors or problems.

SUBTOTAL	$_____

Electrical Supplies/ Hardware

Boxes	$_____
Service entrance panel	$_____
Subpanel	$_____
Circuit breakers	$_____
Fuses	$_____
Connectors	$_____
Wire nuts	$_____
Machine screws	$_____
Wood screws	$_____
SUBTOTAL	$_____

Repair Materials

Caulking compound	$_____
Wood putty	$_____
Sandpaper	$_____
Paint	$_____
SUBTOTAL	$_____
TOTAL	$_____

APPENDIX 3: HOW TO REDUCE YOUR ELECTRICITY BILLS

Take the following steps to keep your electricity bills to a minimum:

■ Install dimmer switches wherever possible for lower energy consumption. (See page 34.)

■ Use fluorescent lighting instead of incandescent lighting; energy savings are substantial (page 45).

■ Install time-clock switches for lighting and appliances—to use energy when you need it and not while you're away from home. (See page 34.)

■ When shopping for appliances, compare wattage ratings and choose the model you prefer with the lowest watts consumption. If the information is presented in amperes, convert to watts using the guide on page 2.

■ Purchase higher wattage light bulbs and use them sparingly rather than using many low wattage bulbs. One 100-watt bulb produces more light than two 60-watt bulbs and yet it consumes approximately 20 percent less energy.

■ Use standard bulbs instead of long-life bulbs everywhere except locations that have difficult access. Standard bulbs take approximately 20 percent less energy to produce the same level of illumination.

■ Switch incandescent lights off when not in use. Although switching shortens a bulb's life slightly, it is a myth that switching consumes energy.

■ Leave fluorescent lights switched on. Frequent switching of fluorescent lights usually consumes more energy than leaving them on for long periods.

■ Purchase lampshades with light materials for good reflective value, shaped with large openings at top and bottom for greater light emission. Position lamps in rooms to reflect light from two walls instead of one.

■ When buying a water heater, check the small on-demand types. Consider insulating your water heater.

■ Carefully check energy ratings for refrigerator-freezer combinations. Generally, upright-designed and automatic-defrost models use more energy than those with chest-type designs and manual defrost. Keep refrigerator coils and vents clean for efficient service. Keep door closed when not in use.

■ When purchasing stoves, consider a convection type oven plus a supplementary microwave oven. Allow foods to thaw before cooking. When baking, put more than one item into the oven for the most efficient use of heat. Use microwave ovens for small quantities of food; cooking for five or more people with a microwave is not energy-efficient.

■ Use a minimum amount of hot water when doing laundry by filling only to the load level and using the coolest temperature setting possible.

■ Clean clothes dryer vent after every use for maximum heat output. Do not overdry clothes. Avoid excessive ironing.

GLOSSARY OF HOME WIRING TERMS

Ampacity. The current in amperes a conductor can carry continuously under the conditions of use without exceeding its temperature rating.

Ampere. A unit of measure of electrical current flow.

Armored cable. Cable sheathed in flexible metal armor. Also called BX cable.

AWG. The standard system for measuring wire in the United States; the American Wire Gauge system.

Back-wired. Description of switches and receptacles that have push-in type terminals on their backs.

Ballast. A device in fluorescent lighting fixtures that starts and stabilizes illumination.

Bare wire. A wire with no outer covering, often used as a ground wire.

Box. A device, metal or plastic, used to house wire terminations and connections at outlets and junctions.

Branch circuit. A circuit running from the service entrance panel to one or more outlets.

Bus bar. A solid conductor at the service entrance panel to which neutral and/or ground wires of branch circuits are connected.

BX cable. A trade name; see *Armored cable.*

Cable. A conductor composed of two or more wires surrounded by a plastic or metal sheathing.

Cartridge fuse. Cylindrically shaped fuses of a higher voltage than plug fuses, used at the service panel or for individual appliance circuits.

Circuit. The closed pathway of electrical current beginning at the source of supply, traveling to outlets, and back again to the supply source.

Circuit breaker. A device that breaks open the circuit when the circuit exceeds a given amount, thereby shutting down the power; can be reset.

Color coding. A standardized system of using wires with certain colors to identify their function.

Common terminal. The hot terminal on a three-way switch.

Conductor. Any material with minimal resistance to electricity—such as copper wire or metal bars. See *Wire.*

Conduit. Tubing of metal or nonmetallic material for enclosing electrical wire.

Connecting block. In telephone wiring, one of the three types of devices used to terminate wires entering the home; also called a *42A block.*

Continuity tester. A testing device with its own power source, used to check faulty wiring.

Conversion jack. In telephone wiring, a device mounted onto a connecting block which converts it to a modular jack; also called an *instant jack.*

Device box. A box housing a switch or receptacle.

Double-pole switch. A switch used to control a single major appliance; it has four terminals.

Duplex receptacle. A receptacle wired for two separate outlets.

EMT conduit. A type of metallic conduit; also called *thinwall conduit.*

End-of-the-run. An outlet that is the last on a branch circuit; only one cable enters an end-of-the-run box.

Equipment grounding conductor. A bare or green insulated conductor found within type NM cables. In this book referred to as the *grounding wire.*

Ferrule-type fuse. A type of cartridge fuse.

Fish tape. A stiff, flexible, metal tape with a hooked end, used to pull wires through inaccessible areas.

Four-way switch. A switch with four terminals that operates with other switches to control power from three or more locations.

Fuse. A device that breaks open the circuit, thereby shutting down the power when the circuit exceeds a given amount; metal in the fuse melts when this happens and the fuse must be replaced.

Ganging. A method of joining two boxes together.

General purpose circuit. A branch circuit carrying the smallest load in the home; sometimes called a *lighting circuit.*

Ground Fault Circuit Interrupter. A very sensitive circuit-breaking device used outdoors and in any damp areas of the home. Also called a *GFCI.*

Grounding. Connecting the components of the electrical system and directing the current to the ground for safety.

Grounding electrode systems. Several relatively new methods of grounding that take into account the use of nonmetallic plumbing pipes and fittings, thus making old methods inadequate. Required for new buildings; recommended when upgrading old wiring.

Handy box. An outlet box with a cover used in surface wiring. Also called a *utility box.*

Hot wire. A cable or wire that carries voltage.

Individual appliance circuit. A branch circuit with a large load needed for an individual major appliance.

Internal clamp. An accessory that fits inside a metal box to clamp cable in place. Also called a *saddle clamp.*

Knife-blade fuse. A type of cartridge fuse most often used at the service entrance panel for the main connection between incoming power and branch circuits.

Knob and tube wiring. An outdated form of home wiring consisting of porcelain tubes through which wire was run and knobs by which wire was supported.

Knockouts. Scored concentric or eccentric sections of metal boxes that, when pushed open, provide access for cable.

Leads. Wires leading from fixtures such as special switches and lighting fixtures.

Live wire. See *hot wire.*

Low voltage. A current flow pressure much lower than the standard 120 volts in homes—20 to 30 volts or less.

Middle-of-the-run. Wiring within a box that contains two cables.

Modular interface. In telephone wiring, one of the three types of devices used to terminate wires entering the home; also called an *SNI.*

Modular outlet. In telephone wiring, one of the three types of devices used to terminate wires entering the home.

National Electrical Code. The body of regulations sponsored by the National Fire Protection Association which serves to protect from dangers due to the use of electricity. All local codes are based on the *National Electrical Code.* Also called the *NEC* or the *Code.*

Neutral wire. A conductor in cable that carries return or 'dead' current throughout the circuit.

NM cable. A type of nonmetallic sheathed cable; NM stands for 'nonmetallic'.

NMC cable. A type of nonmetallic sheathed cable that is waterproof.

Open circuit. A broken electrical circuit through which no current can flow.

Outlet. Any point in the wiring system where current supplies usable equipment. Lighting fixtures, switches, and receptacles are all outlets.

Overcurrent device. A device, such as a fuse or circuit breaker, that is used to prevent an excessive current flow.

Overload. The condition whereby the current demands more energy than the circuit or the equipment can withstand.

Pigtail. A short wire spliced to two or more wires and connected by its other end to a terminal; used because only one wire may be connected to a terminal.

Plug fuse. A device containing a metal strip that melts when overheated—a condition caused by an overload or a short circuit.

Polarized. Referring to plugs and receptacles—matched sizes of prongs and plugs for safety.

Pulse dialing. A system of telephone dialing which sends on-and-off signals to the telephone company; also called *rotary dialing.*

PVC. A type of nonmetallic conduit, polyvinylchloride, sometimes used in outdoor wiring.

Quick-clamp. A feature of some plastic boxes that permits inserting cable and clamping it without using a screw.

Raceway. A wiring system of conductors and channels installed on wall surfaces.

Resistance. The ability of a material to resist the flow of current. Poor conductors like wood and glass have much resistance.

Rigid metal conduit. A type of metal conduit sometimes used in outdoor wiring; also called *heavywall conduit*.

Ring conductor. In telephone wiring, one of the conductors—usually red and identified as *R*.

Romex. A trade name of nonmetallic sheathed cable.

Service Entrance Panel. Technically, the *main service equipment panel board*. Location from where service equipment is housed, grounded, and distributed throughout the home via branch circuits. Also called the *panelboard, source,* or *main*.

Service head. The point at which utility company wires are attached to a house before entry.

Short circuit. An incorrect connection between two hot wires or between a hot wire and a neutral wire.

Side-wired. Description of switches and receptacles that have screw-type terminals on their sides.

Single-pole switch. The most common kind of switch, with two terminals, that controls a switch or receptacle from one location.

Small appliance circuit. Description of a branch circuit with a load larger than a general purpose load but much smaller than an individual appliance load.

Splicing. The process of connecting two or more wires.

Split-wired receptacle. A dual receptacle in which each receptacle is connected to a different branch circuit.

Station protector. In telephone wiring, a device located outside the home; the responsibility of the telephone company.

Stripping. The process of removing sheathing from wiring.

Surface wiring. Wiring that is located on the surface of walls and ceilings rather than concealed behind them.

Switch loop. A circuit run in which a lighting fixture is installed between a power source and a switch.

T type wire. A type of wire with thermoplastic sheathing.

TW type wire. Wire with a weatherproof sheathing; the type most often used in home wiring.

Terminal block. In telephone wiring, an installation within the home where telephone company wiring terminates.

Three-way switch. A switch with three terminals that operates with another switch to control power from two locations.

Time-delay fuse. A type of screw-in fuse which melts when there is a short circuit but does not when there is a momentary overload, such as one caused by the brief starting of a motor.

Tip conductor. In telephone wiring, one of the conductors—usually green and identified as *G*.

Tone dialing. A modern and widely used system of telephone dialing which sends combination tones to the central office to signal the number called.

Track lighting. A type of lighting which incorporates surface wiring and fixtures that are 'plugged into' a track.

Transformer. A device used in low-voltage wiring to step down the power from a home circuit to the low voltage required.

Traveler terminal. A terminal on a three-way switch that transfers the current from one switch to another.

Two-part clamp. A fitting used in metal boxes to clamp cable to the box; consists of two parts, a threaded bushing and a locknut. Also called a *locknut connector*.

Type S fuse. A two-part fuse with sized threading, designed to ensure usage of the correct size fuse. Also called a *nontamperable fuse*.

UF cable. A type of waterproof nonmetallic-sheathed cable that is especially designed for outdoor use. UF stands for 'underground feeder'.

UL. Abbreviation for *Underwriters' Laboratories*, a nonprofit organization that tests electrical equipment in accordance with safety standards. Listed products are marked with a UL symbol.

Underwriter's knot. A sturdy knot used in wiring plugs and lamps.

Volt. A unit of measure of electrical force or pressure.

Voltage tester. A testing device used to check for voltage at an outlet—to make sure that the power is turned off.

Watt. A unit of measure of electrical power. VOLTS X AMPS = WATTS.

Wire. A strand or several strands of highly conductive material insulated by a protective covering.

Wire nut. A plastic cap-like fixture for splicing wires; the interior is composed of threaded metal.

INDEX